Mobile Robotics
for Multidisciplinary Study

Synthesis Lectures on Control and Mechatronics

Editors
Chaouki Abdallah, *University of New Mexico*
Mark W. Spong, *University of Texas at Dallas*

Mobile Robotics for Multidisciplinary Study

Carlotta A. Berry

ISBN: 978-3-031-00702-6 paperback
ISBN: 978-3-031-01830-5 ebook

DOI 10.1007/978-3-031-01830-5

A Publication in the Springer series
SYNTHESIS LECTURES ON CONTROL AND MECHATRONICS

Lecture #4
Series Editors: Chaoiki Abdallah, *University of New Mexico*
 Mark W. Spong, *University of Texas at Dallas*
Series ISSN
Synthesis Lectures on Control and Mechatronics
Print 1939-0564 Electronic 1939-0572

Mobile Robotics
for Multidisciplinary Study

Carlotta A. Berry
Rose-Hulman Institute of Technology

SYNTHESIS LECTURES ON CONTROL AND MECHATRONICS #4

ABSTRACT

This lecture series provides an introduction to the field of mobile robotics and the intersection between multiple robotics-related disciplines including electrical, mechanical, computer, software engineering and computer science. It is intended for an upper-level undergraduate or first-year graduate students interested in mobile robotics and artificial intelligence with some experience in object-oriented programming and controls. Focus areas will include robotics history, hardware, control and software. Specific topics include robot components, effectors and actuators, locomotion, kinematics, sensors, feedback control, control architectures, representation, navigation, localization and mapping. The end of each chapter includes review questions as well as exercises to provide applications for the concepts as well as opportunities for further study.

KEYWORDS

mobile robotics, artificial intelligence, control architectures, behaviors, navigation, localization, mapping

Contents

Preface

My motivation for the creation of this lecture series was to address a dearth in mobile robotics textbooks that were appropriate for my needs. I wanted a textbook that was designed for upper-level undergraduates or first-year graduate students in multiple disciplines. Ideally, this text would not be discipline specific, platform specific or language specific. In addition, the book would focus on the entire spectrum and state of the art as opposed to solely hardware or software. It would survey topics related to robotics history, theory, control and applications, and would provide a summary of the state of the art at a level that the novice roboticist could understand. It would not only provide enlightenment but also inspiration.

This lecture series contains topics on mobile robotics history, hardware, theory, control, software and application. My vision is that after reading each chapter, the review questions and exercises will provide a roadmap that the reader can use to replicate the concepts on their robot. Then, after experiencing some success, the reader will be motivated to study the topics in more depth and begin to make their own mark on the robotics field.

Carlotta A. Berry
March 2012

Acknowledgments

First and foremost, I would like to acknowledge God for guiding my life toward electrical engineering, controls, robotics and education. I know that to whom much is given, much is required and I will continue to strive to do more and give more (Luke 12:48). I would like to also thank my mother for always being my biggest cheerleader and instilling in me a great appreciation for learning along with an intense work ethic. In addition, none of this would have been possible without the unwavering support of my husband and I thank him every day for being patient with me as I logged long hours reading and writing. Also, to my little girl, I know that "mommy has to work" can be hard for you to understand but I promise that everything I do is for my family and the best is yet to come! I love you all very much.

I would also like to acknowledge my colleagues in robotics at Rose-Hulman; who would have ever imagined that our vision for innovation would blossom into something so great? I also would like to thank the Rose-Hulman Electrical and Computer Engineering department and corporate sponsors for allowing me to be me. I also thank my robotics students who challenged me, taught me and always pushed me to learn more and do more. Lastly, I want to acknowledge some key figures in robotics whose work was essential in my study and the completion of this lecture series including Julie Adams, Kazuhiko Kawamura,Robin Murphy, Maja Matarić, Ron Arkin, Rodney Brooks, Roland Siegwart and Illah Nourbakhsh.

Carlotta A. Berry
March 2012

CHAPTER 1

Introduction

1.1 MOTIVATION

Why study *robotics*? Well, simply put, *robots* inspire imagination, innovation and invention. **Robots** are mechanical systems with electrical controls given intelligence through software [1]. Thus, by definition, a diversity of disciplines study, design, build and control robots. As such, the field of robotics is ideal for illustrating the intersection between fields such as biomedical, computer, electrical, mechanical, software engineering and computer science. **Robotics** is the study of autonomous systems operating in the physical world in terms of perception, planning and action. In the last 30 years, robotics has become very popular for people of all ages and proven to be a great draw for science and engineering. Thus, this lecture series will hopefully inspire and motivate the reader to learn and do more mobile robotics by presenting a summary of the history and current state of the art with applications.

1.2 ROBOTICS FOUNDATIONS

In order to inspire innovation in robotics, the first step is to look at the history of what has already been done. This survey of the land provides context for the current state of the art. The term "robot" was first used in 1923 by a Czechoslovakian playwright, Karel Capek, in his play, R.U.R. (Rossum's Universal Robots). He used it to describe a set of servant machines or menial laborers who attacked the humans when they were given emotions. This perspective of the robot as a mindless laborer would motivate the first incarnations of an actual robotic system. In 1942, during World War II, robotic-type systems were used to handle radioactive materials to protect humans as they built nuclear bombs. In addition, artificial sensing and autonomous control precursors were used for radar tracking the enemy as well as mine detectors.

Isaac Asimov provided one of the first perspectives of a robot as an autonomous anthropomorphic mechanical system in his science fiction collection, "I, Robot." In this series, robots had some logic and reasoning but still did not always perform as expected. A key product of his work were the three laws of robotics.

1. A robot must not harm a human being or cause one to be hurt through inaction.

2. A robot must obey a human beings orders unless it contradicts law one.

3. A robot must protect its own existence unless it contradicts laws one or two.

A contemporary of Asimov was Norbert Wiener, a scientist who invented cybernetics and control systems. **Control theory** originated in Greece and is the study of how to design or control machines to track or regulate to a certain set point. **Cybernetics** is a combination of the field of biological systems, controls and communication theory. Cybernetics examines common properties between animals and machines. His work was the beginning of using artificial devices to integrate sensing, processing and action within an environment. Cybernetics would eventually branch off into two distinct fields of research: artificial intelligence and robotics.

In the 1950s, W. Grey Walter, a British neurophysiologist, built what is considered to be the first fully autonomous robot. His pair of robots, Elmer and Elsie, were referred to as Tortoises. These robots had analog electronics via vacuum tubes, microphone ears and contact switch feelers. They could avoid obstacles, find light move toward or away from light, and move to a battery recharging station as needed. This combination of actions made these robots have animal like behavior (attraction and repelling) that simulated biological systems.

The term "Artificial Intelligence (AI)" was defined in 1955 at the DARPA summer research conference at Dartmouth University attended by Marvin Minsky, Allan Newell, John McCarthy and Herbert Simon. **Artificial Intelligence** is the science of making a machine more intelligent by creating its own abstract model of the environment. AI has several competencies including:

- Knowledge representation - Internal models of the world

- Understanding natural language

- Learning

- Planning and reasoning to solve problems

- Inference

- Search through possible solutions

- Vision

- Symbolic representation of information

- Hierarchical system organization

- Sequential program execution

It is evident from this list that a robot's intelligence is strongly tied to the speed and amount of information it can sense about its environment and how it can use this information to achieve some goal or perform some task.

In the 1960s, robot manipulators and automation changed the face of industrial manufacturing. Industrial manipulators typically have a shoulder, elbow and wrist and no mobility since they are affixed to the factory floor. However, they do have very well-defined dynamic and kinematic models

so that they can be preprogrammed to complete operations such as picking and placing parts or welding at a high level of accuracy, precision and repeatability. These systems typically had no sensor feedback, logic or cognition, i.e., open loop control.

Shakey, developed at Stanford Research Institute (STRI) in the 1970s, was considered to be the first AI-inspired robot. It was so named because it would often shake when it executed a path. This robot was five feet tall and had stepper motors, contact bumpers, optical range finder and camera. It was housed in an artificial world with a flat white floor and black obstacles such as blocks. It used automated reasoning and planning software called STRIPS (STRI Problem Solver) to find a box of a given shape, size or color, construct a map of the world, plan a path, move to the box and push it to a given location. The downfall of this system was that, due to the state of processing power at the time, it was very slow. At the same time that Shakey was being developed, the NASA space program was also working on using AI techniques to build planetary rovers to go to the moon. The application of AI techniques to robotics created the following abbreviated list of AI competencies:

- Learning

- Planning

- Reasoning

- Problem solving

- Knowledge representation

- Computer vision

To this day, it is still very difficult to design any one robot that exhibits all of these features. However, in order to have intelligence it must contain some of them. The key difference between industrial robots and AI robots is that industrial robots were based upon control theory that included very accurate environment, dynamic and kinematic models while AI robots were designed for unpredictable events in an unstructured world [2]. However, it is clear that as manufacturing applications become customizable and require more flexibility; AI techniques will be required to design industrial manipulators.

In the late 1970s, Hans Moravec of the Stanford University Artificial Intelligence Lab designed the Stanford Cart. This robot used stereovision to find objects and plan paths although not much faster than Shakey. It was then moved to Carnegie Mellon (CMU) in 1980 and the CMU Rover was designed. It improved upon Shakey because it used synchronous drive instead of Ackerman steering. In addition, the CMU Rover also included IR and sonar as well as a camera with pan and tilt instead of on a sliding track. Another contemporary of Shakey was HILARE developed at the LAAS lab in Toulouse, France; it had a video camera, ultrasound sensors and laser range finder. This robot accomplished another robotics milestone by being an indoor robot based on a differential drive system. It generated models of the world as geometric shapes and used them to navigate.

In 1984, Valentino Braitenberg, a psychologist, wrote a book entitled "Vehicles" about a set of vehicles that were made incrementally more complex by coupling sensing to motors. These systems had excitatory and inhibitory connections which created photophilic and photophobic behaviors. Thus, by the early 1980s it was becoming evident that there was a need for robots that could think and move faster. This would lead to the divergence from the purely deliberative control seen on Shakey to other control architectures (i.e., reactive, hybrid).

In 1986, Rodney Brooks at MIT created the subsumption architecture which was a parallel arbitration between layers of task-achieving modules. This architecture tightly coupled sensing and acting to create a robot that responded quickly to changes in the environment. Brooks' work later became known as behavior-based control or a behavior-based architecture. Then, Ron Arkin and Michael Arbib created the potential fields and motor-schema based navigation for a mobile robot. Behavior-based control will be explored in more detail later in this text.

In conclusion, this has been only a brief summary of the beginning history of robotics and there are many other milestones that could be explored such as the advent of social (entertainment) robots, service robots, androids, humanoids, multirobot systems, surgical robots, microrobots, nanorobots, human-robot interaction, and the list goes on and on. Also evident from this brief history, the field of robotics is rapidly changing in form and functionality and will continue to do so as more is discovered about it and other technologies. Figure 1.1 provides a compact history of the first 80 years in robotics.

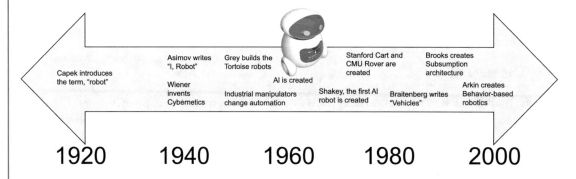

Figure 1.1: History of robotics timeline.

1.3 DEFINITION OF A ROBOT

There are many definitions of a robot but for the remainder of this text, there will be two ruling definitions provided by Robin Murphy and Maja Matarić:

"An intelligent robot is a mechanical creature which can function autonomously [2]".

"An autonomous system which exists in the physical world can sense its environment and can act on it to achieve some goals [3]".

From these definitions, it is evident that the key difference between a mobile robot and a radio-controlled car is autonomy or the ability to couple sensing and planning to act on the world. In addition the key difference between a web bot and a mobile robot is embodiment.

Robots are ideal for tasks that are dull, dirty, dangerous or desolate and the level of autonomy may vary based upon context. *Teleoperation* is when a human operator controls the robot and performs localization and cognition as the robot executes motion and this system has no autonomy. A robot may be semi-autonomous or fully autonomous based upon the level of interaction with a human. In *semi-autonomous control*, the human controls the robot at certain times dependent upon the task. There are two types of semi-autonomous control: supervisory and shared. In *supervisory control*, the human may assign some tasks to the robot to complete autonomously and only observe the progress. In *shared control*, the human assigns tasks to the robot but may interrupt with input such as perception, additional instructions or to cancel execution. In *fully autonomous* control, a robot performs its own perception, planning and acting independently without human input. There are three questions an autonomous mobile robot must answer:

- Where am I? (perception)

- Where am I going? What am I doing? (planning)

- How do I get there? How do I achieve it? (acting)

There are currently several research areas in robotics including, but not limited to:

- Wheeled mobile robots

- Legged mobile robots

- Manipulators

- Industrial automation

- Flying robots

- Underwater robots

- Robot vision

- Artificial intelligence

This text will focus on wheeled mobile robots and how to program them to create some level of intelligence or autonomy. Mobile robots are essentially different from other robots because they deal with the problem of sensing, planning, navigating and achieving tasks in large-scale potentially unstructured environments [4]. In order to achieve this overarching goal of robot intelligence, the robotic system will be examined in three phases: hardware, control and software. A high-level summary of the organization of this lecture series is shown in Figure 1.2.

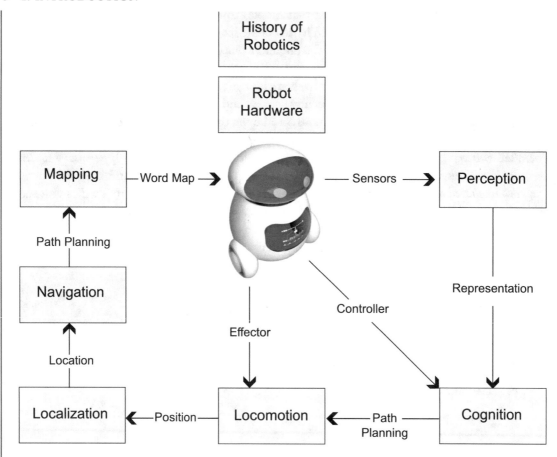

Figure 1.2: Mobile robotics cycle.

Although the diagram contains arrows and appears to be cyclical, some of these processes may happen simultaneously, in a different order or not at all. For example, localization and mapping may happen simultaneously (i.e., SLAM) or sensing and acting may happen in parallel with no cognition (i.e., reactive control).

1.4 REVIEW QUESTIONS

This chapter presented a brief history of robotics as well as a summary of some of the key definitions in the field. Each chapter will conclude with a set of review questions to highlight the key concepts presented in the chapter.

1. In your own words, what is a robot?

2. In your own words, what is robotics?

3. In your own words, what is cybernetics and how does it relate to robotics?

4. What was the first autonomous robot and what features did it have?

5. Describe three characteristics of artificial intelligence and how they relate to robotics theory?

6. What was the first AI-inspired robot and what features did it have? How do these features relate to the key competencies in artificial intelligence?

7. What is one key difference between industrial manipulators and AI robots?

8. What is one key difference between industrial manipulators and mobile robots?

9. What was the motivation for the development of the subsumption architecture and behavior-based robotics?

10. Describe a robotics task that would involve autonomous or semi-autonomous control and why.

1.5 EXERCISES

1. Search the internet for two different robotics research labs and describe their field of research and the platforms used. For a start, you might try Carnegie Mellon, Massachusetts Institute of Technology, Stanford University, NASA Jet Propulsion Laboratory, Virginia Tech and Georgia Institute of Technology.

2. Select one of the following key individuals in robotics history and perform an internet search on their name. Describe the type of robotics related content available on the web that enhances what you have learned about the history of robotics, research and/or applications.

 • Karel Capek

 • Norbert Weiner

 • W. Grey Walter

 • Marvin Minsky

 • Valentino Braitenberg

 • Hans Moravec

 • Rodney Brooks

 • Ron Arkin

 • Maja Matarić

8 1. INTRODUCTION

- Cynthia Brazeal

- Allan Newell

CHAPTER 2

Hardware

2.1 ROBOT PARTS

One objective of this lecture series is to explore mobile robot intelligence. In working toward that goal; it is necessary to examine the primary components of a robotic system. Since a robot is an autonomous system that must sense, plan and act; it must have sensors for perception, a controller for planning and effectors that act on the world. As highlighted by the AI competencies, it is evident that a robot's intelligence is closely tied to the quality and quantity of its sensors, how quickly it can process the data from the sensors and the ability to use that information to act on the world. Figure 2.1 provides a summary of the key components of a robot.

Figure 2.1: Robot parts.

2.1.1 CONTROLLERS

The *controller* is the brain of the robot and performs the cognitive functions. A controller uses a sensor percept to determine how the robot should act upon the world: it is the coordinator. A robot's controller is typically a microcontroller or a computer. A microcontroller is an inexpensive single chip computer with a CPU, I/O, RAM, ROM, timers, parallel/serial ports, A/D and D/A converters. A microcontroller has a preprogrammed output based upon an electrical input; it can store, upload and run a program. Simple robotic systems are controlled by a single central processor. Complicated robotics systems may have basic low-level control hardwired and high-level functionality programmed. In addition, complicated systems may have distributed control where some communication is via wireless, Bluetooth or radio modem. There are tradeoffs between these

two options because on-board computation may be limited by weight and power consumption but offline processors may have bandwidth limitations.

2.1.2 SENSORS

Sensors are used to acquire data from the robot or its environment. Robots require sensors for navigation, localization and mapping. Typically, a *transducer* must be used to convert the measured quantity (sound, distance, temperature) to an electrical quantity (analog, digital) that the controller can use. Sensors may measure a feature and convert it to a resistance, current or voltage (i.e., photocell, phototransistor, infrared). A digital sensor has a binary output such as on and off (i.e., touch sensor). An analog sensor has a continuous output of values such as a range of measurements (i.e., sonar). A PWM (pulse width modulation) sensor pulses a train of measurements between 0 and 5 volts (i.e., modulated infrared) to encode a signal to minimize interference.

 Perception is when the robot uses the sensor information to extract useful information about the robot's internal or external state. This model of the robot's world (internal and external) is *representation*. Figure 2.2 provides a model of robot sensing.

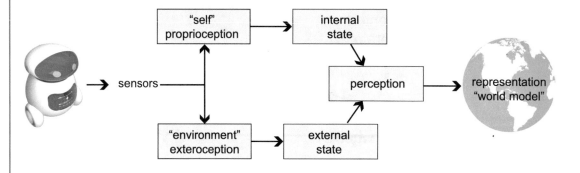

Figure 2.2: Sensing scheme.

Categories and Types
There are several types of sensors that measure a robot's internal and external state. *Exteroceptive* sensors measure the robot's environment such as light, temperature, distance and sound. *Proprioceptive* sensors measure the robot's internal state such as battery power, motor load, wheel position or velocity. The combination of these two sensors is the robot's *perceptual system* and the *sensor space* is the combination of all of the potential sensor readings.

 Sensors can also be divided based upon the sensing mechanism and the level of processing. A *passive* sensor uses a detector to measure energy from the environment (i.e., camera, microphone). An *active* sensor uses an emitter to send some type of energy (light, sound) to the environment and a detector to measure the quantity returned (i.e., sonar, infrared, laser). A *simple* sensor does

not require a great deal of processing to extract information (i.e., distance to objects, light level). A *complex* sensor requires a great deal of processing of data (i.e., feature extraction). Table 2.1 provides a categorization of some of the more popular sensors used in mobile robotics.

Table 2.1: Sensor categories		
	Passive	**Active**
Simple	Bumpers	Infrared
	Compass	Encoder
	Gyroscope	
Complex	Camera	Sonar
		Laser

Sensors can also be classified by what they measure such as proximity, range, motion/speed, position or orientation. Table 2.2 provides a classification of various sensors by their use.

Table 2.2: Sensor classifications		
Classification	**Sensor**	**Details**
Heading	Gyroscope	uses angular acceleration to measure orientation
Heading	Compass	uses the earth's magnetic field to measure heading
Motion/Speed	Encoder	measures motor shaft turns to estimate position or velocity
Motion/Speed	Doppler radar	transmits and measures electromagnetic or sound waves to estimate velocity
Position	GPS	uses a constellation of satellites to estimate absolute position
Proximity	Bumper	uses a tactile switch to indicate the presence of an object
Range	Infrared	emits and measures reflected infrared light to determine distance to objects
Range	Sonar	emits a ping and measures sound from an object to estimate distance to objects
Range	Laser	emits highly amplified and coherent radiation and measures range using time of flight

Sensor Diversity

Sensors are prone to errors, bad readings, limited range and resolution, thus it is beneficial to have sensor diversity. Robots should have different types of sensors to allow for flexibility in the types of data and to provide a more accurate representation of the world. Multiple sensors may speed up the rate of environment readings, save power and provide redundancy and fault tolerance. The types of

sensor diversity may include measurements form different sensors, different positions or at different times.

Sensor arbitration is a means of combining multiple sensors to get better information about the world. There are three types of sensor arbitration: competing, complementary and coordinated [2]. Sensors that return the same percept are redundant or *competing* such as a robot with two sonar rings. Sensors that are logically redundant return identical percepts but with different modalities or levels of processing such as infrared and sonar. This is also referred to as *sensor fission*. *Complementary* sensors acquire disjoint information about the same percept. This is also referred to as *sensor fusion*. For example, if the robot must find a red ball; the camera may be used to identify color, infrared to find location and laser to identify size and shape. *Coordinated* sensors use a sequence of sensors to acquire a percept such as firing individual sonar on sonar ring at different times. This is referred to as *sensor fashion*.

There are several research techniques for actual data fusion which involves combining data from multiple positions, times or sensors. These methods include interpolation, extrapolation and finding an analytic function to fit the data. This estimate from the combined data typically also has an estimate of the confidence of the final estimate given the estimate of the individual readings. The mathematical computation for the data fusion is based upon least squares combination of estimates or Kalman filtering, Monte Carlo, Markov or condensation techniques [4]. Some of these algorithms will be discussed in more detail in Chapter 4.

2.1.3 EFFECTORS AND ACTUATORS

An *effector* is any device that has an effect on the environment (i.e., wheels, tracks, arms, grippers). An *actuator* is the mechanism that enables an action or movement (i.e., electric motors, solenoids, hydraulic or pneumatic cylinders). The most popular actuator for robots is a DC motor because it is inexpensive, small in size and has a high energy output. It is possible to use PWM to drive the motor at variable speeds. Stepper motors are used for applications that require high accuracy in position. In order to create high position accuracy with a DC motor it must be converted to a servo motor by adding sensor feedback via an encoder to measure wheel position. A servo motor may also have a control circuit such as PID controller to adjust the wheel to minimize error between the desired and actual position. The key differences between DC and stepper motors is the accuracy and that the DC motor has high torque at low speeds and the stepper motor has the highest torque at high speeds. Stepper motors typically use a simpler open loop control as opposed to closed loop control for servo motors. Closed loop control and PID control will be discussed in more detail in Chapter 3.

2.2 LOCOMOTION

Locomotion refers to the way that a robot moves from place to place. In locomotion, the environment is fixed and the robot moves by imparting force on the environment. There are two primary concerns in locomotion:

- move the robot to a particular goal location;

- move the robot along a trajectory.

Navigation (path planning) is concerned with getting the robot to a goal and this will be described in more detail in Chapter 4. *Trajectory (motion) planning* is a little bit more difficult than navigation because it deals with following a path where the robot may need to find or follow the shortest, safest or more efficient path with a time constraint.

 Odometry or path integration is a means of implementing dead reckoning to determine a robot's position based upon the robot's prior position and the current heading and velocity. The advantages of this method are that it is self contained, it always provides an estimate of position, and the position can be found anywhere along curved paths. The disadvantages are that the position error grows and require accurate measurement of wheel velocities over time. There are several types of odometry error including systematic from unequal wheel diameters, misalignment of wheels, finite encoder resolution and finite encoder sampling rate. There is also non-systematic odometry error such as travel over uneven floors, unexpected objects on the floor and wheel slippage. Lastly, there are odometry errors such as imprecise measurements, inaccurate control models and immeasurable physical characteristics. These inaccuracies in dead reckoning cause a robot traversing a square path to yield a result similar to Figure 2.3.

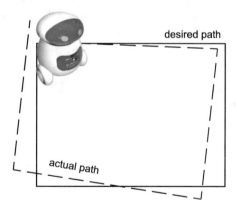

Figure 2.3: Robot odometry error.

 Thus, dead reckoning is most appropriate for short distances because of the error accumulation. For accuracy, an encoder or some other feedback sensor should be used to periodically null out or reset the accumulation error. Another method for correcting odometry error is to use localization which will be discussed in Chapter 4.

 Borenstein et al. developed a method to quantitatively measure the odometry error in a wheeled mobile robot [5, 6]. This method enables the user to calibrate the system based upon systematic errors. The tuning factors found using this technique can be incorporated into the software to reduce

systematic error. This method is called the University of Michigan Benchmark (UMBmark). The robot is programmed to traverse a square path of 4 x 4 m length and four 90° spins. The robot traverses the square five times clockwise (cw) and five times in the counterclockwise (ccw) direction. After each run, the difference in the final x and y position between the actual return position and the one computed by odometry is calculated. If this is plotted on a graph, there will be two distinct quadrants, one for the cw runs and one for the ccw runs. The center of gravity for the cw and ccw runs can be found from

$$X_{c.g.,cw/ccw} = \frac{1}{n} \sum_{i=1}^{n} ex_{i.cw/ccw} \tag{2.1}$$

$$Y_{c.g.,cw/ccw} = \frac{1}{n} \sum_{i=1}^{n} ey_{i.cw/ccw}, \tag{2.2}$$

where n is the number of times the robot runs clockwise or counterclockwise runs ($ex_{i,cw/ccw}$, $ey_{i,cw/ccw}$) is the offset of the final position in Cartesian coordinates from the initial position in Cartesian coordinates. The distribution of the readings within each quadrant is a result of nonsystematic errors. However, it is still evident that the contribution due to systematic errors is much larger than the non-systematic errors. The asymmetry in the centers of gravity between the cw and ccw runs is due to two types of systematic errors: Types A and B. Type A errors are due to an increase or decrease in robot rotation for both cw and ccw runs. Type B errors are due to an increase or decrease in robot rotation for the cw run and the opposite effect happens when the robot travels in the ccw direction. Type A errors occur because of uncertainty in the wheelbase measurement while Type B errors occur because of unequal wheel diameters. After this experiment, a single numeric value that expresses the odometric accuracy for systematic errors can be found from

$$E_{max,syst} = \max(r_{c.g.,cw}; r_{c.g.,ccw}) \tag{2.3}$$

where

$$r_{c.g.,cw} = \sqrt{(x_{c.g.,cw})^2 + (y_{c.g.,cw})^2} \tag{2.4}$$

and

$$r_{c.g.,ccw} = \sqrt{(x_{c.g.,ccw})^2 + (y_{c.g.,ccw})^2} . \tag{2.5}$$

From this calibration, there are two constants that are used in the basic odometry computation. The constants are α and β given by the following formulas assuming the square path has a side of length D, and the robot has a wheel base, B:

$$\alpha = \text{average} \left(\frac{x_{c.g.,cw} + x_{c.g.,ccw}}{-4D}, \frac{y_{c.g.,cw} - y_{c.g.,ccw}}{-4D} \right) \tag{2.6}$$

$$\beta = \text{average} \left(\frac{x_{c.g.,cw} - x_{c.g.,ccw}}{-4D}, \frac{y_{c.g.,cw} + y_{c.g.,ccw}}{-4D} \right) . \tag{2.7}$$

The tuning factors for the wheel base, c_b, left wheel, c_l and right wheel, c_r are

$$E_d = \frac{D + B\sin(\beta/2)}{D - B\sin(\beta/2)} \tag{2.8}$$

$$c_b = \frac{\pi}{\pi - \alpha} \tag{2.9}$$

$$c_l = \frac{2}{E_d + 1} \tag{2.10}$$

$$c_r = E_d c_l \, . \tag{2.11}$$

These values could be used to scale the drive commands such as velocity or distance. This method was found to improve the odometry error by ten- or twenty-fold. The authors also devised a method to improve non-systematic error as well but it will not be addressed in this treatise.

2.3 KINEMATICS

Kinematics is the dynamic model of how a robot behaves. Kinematics is a description of mechanical behavior for design and control. Mobile robot kinematics is used for:

- position or pose estimation (navigation (path planning));

- motion estimation (tracking (trajectory planning)).

The key difference between pose and position estimation is that pose estimation also includes the robot's orientation. Since mobile robots are unbounded with respect to their environment there is no way to directly measure the robot's position, it must be integrated over time which leads to inaccuracies in the estimations. *Forward Kinematics* involves using robot wheel velocities to estimate the robot's motion, position, or orientation at future time. This is the same as the aforementioned dead reckoning and as previously mentioned it is prone to accumulation error. *Inverse kinematics* involves determining wheel velocities in order for a robot to achieve a desired motion, position or orientation. This is a little bit more difficult but there is a simple approach that will be described later in the text.

2.3.1 WHEELED MOBILE ROBOT CLASSIFICATIONS

The four most prevalent wheeled mobile robot classifications are: differential drive, synchronous drive, tricycle drive and car drive (Ackermann or kingpin steering). They are distinguished by the arrangement of the driving and steering wheels. These vehicles may also have one or more casters for stability but those wheels do not impose any kinematic constraints on the motion. Figure 2.4 provides a simplified graphic for each of the primary classifications. There is also a fifth classification, omnidirectional, which will be addressed briefly.

Figure 2.4: Wheeled mobile robot classifications.

Differential drive is the most common type of wheeled mobile robot. It is one of the least complicated with respect to programming and locomotion [8]. It has two wheels driven independently on a common axis and one or two casters. The caster(s) are just for support and do not affect the kinematic model. This configuration allows the robot to drive straight, spin and turn. If the drive wheels run at the same speed, the robot moves forward or in reverse. If the wheels move in opposite directions at the same speed, the robot spins in place. If the robot wheels move at different speeds, the robot turns or moves along a curve. *Note that differential drive robots are very sensitive to differences in wheel velocities, thus it is hard to make the robot drive in a straight line!* Some of the disadvantages of differential drive are that it suffers from lateral movement or wheel slippage which is a problem for using odometry to estimate position. In addition, these robots may experience tip over dependent upon the momentum and center of gravity.

In ***synchronous (synchro) drive***, all of the wheels steer and drive. All of the wheels point in the same direction at all times. In order to change directions, the robot turns all of the wheels. Thus, the robot may change direction while the chassis never changes direction because all of the wheels rotate on one vertical axis. There are only two motors total; one controls the steering and one sets the speed of the three wheels. Typically, all of the wheels are connected on a network of belts. One advantage for this type of design is that it resolves the possibility of tip over seen in differential drive robots. However, differential drive and synchronous drive robots are sensitive to variations in the ground plane which cause the wheel slippage.

One solution to the wheel slippage problem is to use steered driving such as tricycle or car drive. There are two types of ***tricycle drive*** robots. One design will have two passive rear wheels and

one driving and steering wheel in the front. Alternately, the two rear wheels are the driving wheels and the front wheel is the steering wheel. *Ackermann steer* (car drive) vehicles are very similar to tricycle drive except that there is a link that connects the two front steering wheels. The steering wheels have an extended axis that intersects at a common point thus the steering wheels turn at different angles. The rear and/or front wheels may be the driving wheels. One additional advantage of tricycle and car drive robots is that it is not necessary to monitor velocities to drive straight, it is only necessary to steer straight. However, as will be seen shortly it is more difficult to derive a kinematic model for estimating position or moving to a desired pose. Therefore, these types of robots are not used when the task requires accuracy in position. However, they are appropriate for outdoor tasks that require long distances over uneven terrain.

Lastly, there are the less popular **omnidirectional** vehicles. This robot has no kinematic constraints and is completely steerable in the two-dimensional plane. This type of robot requires one of two different wheel configurations: Swedish, spherical. Swedish wheels do not have a vertical axis of rotation but there are rollers attached to the standard fixed wheel antiparallel to the main axis. In order to move the robot sideways, the primary wheel is locked and the rollers are engaged. The Swedish wheel functions as a normal wheel but provides low resistance in another direction. Conversely, spherical wheels can be actively powered to spin along any direction. The key difference between Swedish wheels and spherical wheels is that they do not put any kinematic constraint on the robot chassis. It does not have a principal axis of rotation and the wheels are mounted radially symmetric around the robot's base.

2.3.2 FORWARD KINEMATICS

Each robot's drive or steering wheel contributes to the robot's motion and imposes constraints on the robot's motion [7]. The constraints must be expressed with respect to the global reference frame $\{\mathbf{X}_G, \mathbf{Y}_G\}$ and the robot's reference frame, $\{\mathbf{X}_R, \mathbf{Y}_R, \theta\}$. The global reference frame is static and has an origin, O. The angular difference between the robot and the global reference frame is θ. The coordinates of a point on the robot's chassis is $\mathbf{P} = (x, y)$. The vector ${}^G\mathbf{P}_R = [x, y, \theta]^T$ describes the location of the robot with respect to the global reference frame. Figure 2.5 illustrates the robot's reference frame and the global reference frame.

The orthogonal rotation matrix, $R(\theta)$, is used to map the robot's velocity in the global reference frame to the robot's (local) reference frame. Conversely, the inverse of the orthogonal rotation matrix can be used to transform the robot's motion in the local reference frame to the global reference frame. The orthogonal rotation matrix and the conversion from the global to the local frame are given in Equations (2.12) and (2.13):

$$R(\theta) = \begin{bmatrix} \cos\theta & \sin\theta & 0 \\ -\sin\theta & \cos\theta & 0 \\ 0 & 0 & 1 \end{bmatrix} \tag{2.12}$$

$$\dot{P}_R = R(\theta){}^G\dot{P}_R = R(\theta)\begin{bmatrix} \dot{x} & \dot{y} & \dot{\theta} \end{bmatrix}^T . \tag{2.13}$$

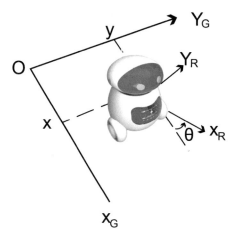

Figure 2.5: Global and robot's reference frame.

Degrees of freedom (DOF) are the minimum number of coordinates that can completely specify the motion of a mechanical system. For example, a helicopter has 6DOF because it can move in 3D space (x, y, z, α[roll], β[pitch], θ[yaw]). A car has 3DOF because it can translate along the plane and rotate or yaw (x, y, θ). A ***controllable degree of freedom (CDOF)*** means that there is an actuator for every DOF. Since the differential drive robot only has control of forward/reverse and rotation (x, θ), it has 2 controllable degrees of freedom. As previously discussed, an omnidirectional robot has 3 controllable degrees of freedom (x, y, θ). Robots with less controllable degrees of freedom than total degrees of freedom are ***nonholonomic***. A nonholonomic robot is restrained in its local movements but not its global movement. A nonholonomic robot will have a complicated trajectory with continuous position and discontinuous velocity to achieve every pose in its space. Differential drive and Ackermann steer robots are nonholonomic. This is evident from the complicated trajectory seen when attempting to parallel park. A ***holonomic*** robot has the same number of CDOF and DOF. An example of a holonomic robot is a quadcopter. A ***redundant*** robot would have more CDOF than DOF. This type of robot may sometimes be seen in manipulators with a shoulder, arm, elbow and wrist.

Differential drive and Ackermann drive robots are nonholonomic, however a differential drive robot has 2CDOF (x, θ) and a car type drive robot has 1 CDOF (x). Differential drive and synchro drive vehicles have a kinematic advantage over Ackermann steer and tricycle drive vehicles because although the trajectory is discontinuous, the differential drive and synchro drive robots can navigate to any position and orientation in the plane. This is not possible with the car-type vehicles because the orientation and position are coupled. However, a differential drive or synchro drive robot can spin to achieve any orientation once it reaches a desired position. In other words, a car-like mobile robot has to move forward or backward to turn while a differential drive robot just spins.

The ***Instantaneous Center of Rotation (ICR)*** is a geometric representation of a robot's kinematic model. For the differential drive robot, it is derived from the drive wheel velocities. For the Ackermann steer drive robot, it is derived from the robot's pose. If a single standard wheel has no lateral movement, then there is a zero motion line through the horizontal axis, perpendicular to the wheel plane. The wheel must move along some circle with a radius, R, the center of this circle is the ICR. For wheels in contact with the ground to roll; they all must move along their own x-axis. Therefore, all of the wheels must trace the circumference of the same circle. The ICR changes over time as a function of the individual wheel velocities and/or the angle of the steering wheels. When R is infinity, the robot moves in a straight line. When R is zero, the robot spins in place. Figure 2.6 presents an example of two different wheel configurations; there is no ICR for the second set of wheels and this robot would not move. The subsequent section will derive the kinematic model for several car types.

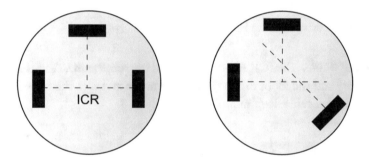

Figure 2.6: Instantaneous center of rotation.

Differential and Synchronous Drive

In this section, the forward kinematic model to describe the motion in the global reference frame will be derived for the differential drive robot shown in Figure 2.7. This model will be used to determine how the robot's wheel velocities affect the overall robot's motion. Note that the robot's angular velocity with respect to the ICR is ω, R is the signed distance from the ICR to the midpoint between the centers of the two wheels, d is the distance along the axle between the center of the two wheels and v_l and v_r are the left and right wheel velocities, respectively. The relationship between the robot's angular velocities and the wheel velocities is shown in (2.14). From these equations, it is possible to solve for the radius of the ICR and the robot's angular velocity as shown in (2.15).

$$v_r = \omega \left(R + \frac{d}{2}, \right), \qquad v_l = \omega \left(R - \frac{d}{2} \right) \tag{2.14}$$

$$R = \frac{d\,(v_l + v_r)}{2\,(v_r - v_l)}, \qquad \omega = \frac{v_r - v_l}{d}. \tag{2.15}$$

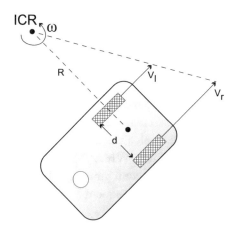

Figure 2.7: Differential drive model.

Note that if $v_l = -v_r$, the ICR radius is zero and the robot spins. Additionally, if $v_l = v_r$, the ICR radius is infinity and the robot drives straight. In order to derive the forward kinematic model with respect to the global reference frame, refer to Figure 2.8.

Figure 2.8: Motion in the global reference frame.

The robot's forward velocity is v and it is the average of the wheel velocities. The robot's angular velocity can be found by finding the contribution of each wheel to the angular velocity. Since each wheel will rotate about the other wheel with radius d, this expression becomes $\omega_r = v_r/d$. The right wheel has a positive rotation and the left wheel has a negative rotations thus the overall angular rotation includes the difference of these values. Next, the inverse of the orthogonal rotation matrix can be used to transform these velocities to be with respect to the global reference frame. This

derivation is shown in Equations (2.16)–(2.19). Recall that differential drive robots have two degrees of freedom since there is no lateral movement, thus the velocity with respect to the y-axis is zero

$$v = \frac{v_l + v_r}{2} \tag{2.16}$$

$$\omega = \frac{v_r - v_l}{d} \tag{2.17}$$

$$\dot{P}_R = \begin{bmatrix} v \\ 0 \\ \omega \end{bmatrix} = \begin{bmatrix} \frac{v_l+v_r}{2} \\ 0 \\ \frac{v_r-v_l}{d} \end{bmatrix} \tag{2.18}$$

$$^G\dot{P}_R = R(\theta)^{-1}\,\dot{P}_R = \begin{bmatrix} \dot{x} \\ \dot{y} \\ \dot{\theta} \end{bmatrix} = \begin{bmatrix} \cos\theta & -\sin\theta & 0 \\ \sin\theta & \cos\theta & 0 \\ 0 & 0 & 1 \end{bmatrix} \begin{bmatrix} \frac{v_l+v_r}{2} \\ 0 \\ \frac{v_r-v_l}{d} \end{bmatrix}. \tag{2.19}$$

Recall that dead reckoning is the process of measuring the distance travelled by a robot by using the odometer. However, the odometer does not measure position or distance from a reference point only relative distance. Therefore, to find the position, the velocity vector must be integrated over the time the motion takes place. For discrete steps, the motion can be expressed as a difference vector and a summation used to find the robot's position. Both of these methods may have acceptable accuracy as long as the steps are small over short distances. An estimate of the robot's position at any time is found by integrating Equation (2.19), given the robot's linear velocity, $V(t)$, in a particular direction, $\theta(t)$, and the robot's initial conditions (see (2.20)–(2.22))

$$x(t) = \int_0^t V(t)\cos(\theta(t))\,dt = \frac{1}{2}\int_0^t (v_r(t) + v_l(t))\cos(\theta(t))\,dt \tag{2.20}$$

$$y(t) = \int_0^t V(t)\sin(\theta(t))\,dt = \frac{1}{2}\int_0^t (v_r(t) + v_l(t))\sin(\theta(t))\,dt \tag{2.21}$$

$$\theta(t) = \int_0^t \omega(t)\,dt = \frac{1}{d}\int_0^t (v_r(t) - v_l(t))\,dt. \tag{2.22}$$

It is also possible to estimate the robot's pose by using the ICR with the center as described in Equation (2.23). If the robot has a pose, $P = (x, y, \theta)$, at some time t, and moves with angular velocity, ω, about the ICR until the time, $t + \delta t$, the ICR is given by Equation (2.23):

$$ICR = (ICR_x, ICR_y) = (x - R\sin(\theta), y + R\cos(\theta)). \tag{2.23}$$

The forward kinematic model to estimate the pose of the robot at $t + \delta t$ in terms of the Cartesian coordinates is given by Equation (2.24):

$$^G\dot{P}_R = R(\theta)^{-1}\dot{P}_R = \begin{bmatrix} \dot{x} \\ \dot{y} \\ \dot{\theta} \end{bmatrix} = \begin{bmatrix} \cos(\omega\delta t) & -\sin(\omega\delta t) \\ \sin(\omega\delta t) & \cos(\omega\delta t) \\ 0 & 0 \end{bmatrix} \begin{bmatrix} x - ICR_x \\ y - ICR_y \\ \theta \end{bmatrix} + \begin{bmatrix} ICR_x \\ ICR_y \\ \omega\delta t \end{bmatrix}. \tag{2.24}$$

In synchronous drive all of the wheels always point the same direction, rotate at the same speed, ω, and move with the same velocity, v. Therefore, a sync drive robot has either an infinite ICR or rotates about its center. The equations of motion for the synchronous drive robot are very similar to those for the differential drive robot and are shown here:

$$x(t) = \int_0^t v(t)\cos(\theta(t))dt \tag{2.25}$$

$$y(t) = \int_0^t v(t)\sin(\theta(t))dt \tag{2.26}$$

$$\theta(t) = \int_0^t \omega(t)dt . \tag{2.27}$$

Tricycle and Ackermann Drive

Many times the kinematics of a mobile robot can be described by a simpler equivalent model to simplify the mathematical process. The equivalent model for a steered drive robot is a bicycle where the two driving wheels are replaced by one wheel at the midpoint of the axle that has an average of the velocity of the two driving wheels. Figure 2.9 illustrates the three equivalent models for the bicycle, tricycle and car steer vehicles.

bicycle drive tricycle drive car drive

Figure 2.9: Steered vehicles equivalent models.

Note that for the Ackermann steer vehicle, the bicycle drive is an approximation because the front wheels turn at a different angle and are mounted on a common axis. Figure 2.10 represents the kinematic model for a bicycle drive robot. The ICR is based upon the robot's pose and must lie at the intersection of an extended axis through the rear wheels and a line perpendicular to the rotational axis of the steer wheel. The pose of the robot with respect to the global reference frame is $(x, y\,\theta)$. The steered front wheel makes an angle of α with the robot's local reference frame. If the steering angle is not zero, then the robot turns with an angular velocity of ω about the ICR with radius, R. In this case, d represents the distance between the front and rear axle of the vehicle.

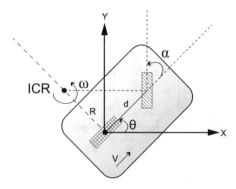

Figure 2.10: Kinematic model for a bicycle.

Based upon the geometry of similar triangles and the Pythagorean theorem, the radius of the ICR and the angular rotation of the robot about the ICR are given by the following equations:

$$R = d \tan\left(\frac{\pi}{2} - \alpha\right), \tag{2.28}$$

$$\omega = \frac{v}{\sqrt{d^2 + R^2}}. \tag{2.29}$$

The forward kinematic model is the same as Equation (2.24) with the appropriate substitution of Equations (2.28) and (2.29).

2.3.3 INVERSE KINEMATICS

Recall that inverse kinematics involves finding the required robot velocity to achieve a certain pose in the robot's environment. There are too many unknowns, not enough equations and multiple solutions. Note that if the robot is nonholonomic, it may not even be possible to achieve a certain pose because of the orientation requirement. Fundamentally, this is a very difficult problem but it can be solved for a few special cases.

Differential and Synchronous Drive

It is possible to solve this problem for a differential drive robot if it is assumed that the left and right wheel velocities are constant and not equal. In this case, the equations of motion reduce to Equations (2.30), (2.31), and (2.32).

$$x(t) = \frac{d}{2}\frac{v_r + v_l}{v_r - v_l} \sin\left(\frac{t}{d}(v_r - v_l)\right) \tag{2.30}$$

$$y(t) = -\frac{d}{2}\frac{v_r + v_l}{v_r - v_l} \cos\left(\frac{t}{d}(v_r - v_l)\right) \tag{2.31}$$

$$\theta(t) = \frac{t}{d}(v_r - v_l). \tag{2.32}$$

If the robot has a pose, $P = (x, y, \theta)$, at some time t, and moves until the time, $t + \delta t$ with constant velocity, $v_l = v_r = v$, the robot moves in a straight line and the equation of motion simplifies to Equation (2.33):

$$\begin{bmatrix} \dot{x} \\ \dot{y} \\ \dot{\theta} \end{bmatrix} = \begin{bmatrix} x + (v\delta t)\cos(\theta) \\ y + (v\delta t)\sin(\theta) \\ \theta \end{bmatrix}. \tag{2.33}$$

Furthermore, if $-v_l = v_r$, the robot spins in place and the equation of motion simplifies to

$$\begin{bmatrix} \dot{x} \\ \dot{y} \\ \dot{\theta} \end{bmatrix} = \begin{bmatrix} x \\ y \\ \theta + \frac{2v\delta t}{d} \end{bmatrix}. \tag{2.34}$$

Therefore, the easiest solution to the inverse kinematics problem is to spin the robot to the desired angle and move forward to the desired location or vice versa. Note that this is not the only solution and there are many possible solutions including some with a smooth, continuous trajectory. These solutions would involve designing a kinematic controller which will not be discussed in this test. To calculate the spin time from the wheel velocities and desired angle use Equation (2.35). To calculate the wheel velocities given the spin time and desired angle, use Equation (2.36). To calculate the forward time if the x coordinate of the pose is the same before and after motion, use Equation (2.37). To calculate the forward time if the x coordinate of the pose is different before and after motion, use Equation (2.38):

$$\delta t = d\frac{\theta(t + \delta t) - \theta(t)}{2v} \tag{2.35}$$

$$v = d\frac{\theta(t + \delta t) - \theta(t)}{2\delta t} \tag{2.36}$$

$$\delta t = \frac{y(t + \delta t) - y(t)}{v\sin(\theta)} \tag{2.37}$$

$$\delta t = \frac{x(t + \delta t) - x(t)}{v\cos(\theta)}. \tag{2.38}$$

Since rotation and translation are also decoupled for a synchronous drive robot, the inverse kinematic model is very similar to the differential drive robot. Therefore, the simplest solution is to perform spin and translation motions independently to navigate to a desired location.

Tricycle and Ackermann Drive

As previously mentioned, the inverse kinematics problem is difficult to solve except for a few limited cases. Therefore, similar to the differential drive and synchronous drive robots it is easier to look at the decoupled case where the robot spins and then drives straight. If the robot is steered straight ($\alpha = 0$), then the kinematic model in Equation (2.24) reduces to

$$\begin{bmatrix} \dot{x} \\ \dot{y} \\ \dot{\theta} \end{bmatrix} = \begin{bmatrix} x + (v\delta t)\cos(\theta) \\ y + (v\delta t)\sin(\theta) \\ \theta \end{bmatrix}. \tag{2.39}$$

The steered robot can actually spin if it is possible to turn the steering wheel $\pm 90°$ and then Equation (2.24) simplifies to

$$\begin{bmatrix} \dot{x} \\ \dot{y} \\ \dot{\theta} \end{bmatrix} = \begin{bmatrix} x \\ y \\ \theta \pm \frac{v\delta t}{d} \end{bmatrix}. \tag{2.40}$$

Note that it is very difficult to design a steered mobile robot that has the ability to execute a pure rotation without some translation [4]. If this steering angle is not achievable, similar to what is seen in a car, it will require very complex trajectories to achieve every possible pose in the robot's space.

2.4 REVIEW QUESTIONS

This chapter has summarized the key components of a robot including the controller, sensors, effectors and actuators. It also describes robot motion in the world including forward and inverse kinematics. The following questions provide a review of some of the content examined in this chapter.

1. What are the key components of a robot and why would the robot need these parts? Can you think of any other part that is integral to being a robot?

2. In your own words, what is the difference between a sensor and a transducer?

3. In your own words, what is the difference between an effector and actuator?

4. Categorize the following list of sensors as proprioceptive or exteroceptive, passive or active, simple or complex: compass, gyroscope, inclinometer, potentiometer, optical encoder, limit switch, sonar, laser, Doppler radar.

5. Give an example of a simple and complex sensor that could be used for sensor redundancy to enable a robot to avoid obstacles.

6. Give an example of an active and passive sensor that could be used for sensor redundancy to enable a robot to follow a wall.

7. Describe one application where it would be more beneficial to use an Ackermann drive or differential drive robot versus another classification.

8. Describe one application where it would be more beneficial to use a holonomic or non-holonomic robot.

9. Is a synchronous drive robot holonomic? Why or why not?

10. Would a camera and infrared sensor be an example of redundant, complementary or coordinated sensors?

11. Would a laser and infrared sensor be examples of redundant, complementary or coordinated sensors?

12. Would a sonar ring be an example of redundant, complementary or coordinated sensors?

2.5 EXERCISES

1. Robot Motion

 (a) Create a random wander routine for a robot that uses a random direction and/or heading in order for a robot to explore its environment.

 (b) Write a program to use odometry to drive the robot in a circle with a diameter of at least 2 ft.

 (c) Create a modular program that uses the circle function in part b to drive the robot in a figure eight.

 (d) Write a program to use odometry to drive the robot in a square with sides of at least 2 ft with 90° turns.

2. Odometry Error

 Replicate the bi-directional square path experiment by Borenstein and Feng at the University of Michigan. You can view details regarding the experiment on p. 134 of the paper, "Where Am I?" available online. What is the odometric accuracy for systematic errors for your robot? Were you able to improve this value by adding the odometry tuning factors?

3. Obstacle Avoidance

 (a) Write a *"scared kid"* obstacle avoidance program where the robot sits still until an object or person comes near (within 5 in). The robot then feels the force vector from the sensors and moves in the opposite direction until all sensors are clear and then stops.

 (b) Write an *"aggressive kid"* obstacle avoidance program where the robot is in a continuous motion such as moving straight, in a circle or square until an object comes near (within 5 in). The robot then feels the force vector from the sensors and moves in the opposite direction until all sensors are clear and continues the motion.

4. Use MATLAB to simulate the path of a differential drive robot by using the ICR to represent the kinematic model.

CHAPTER 3

Control

Now that we are familiar with the key components of a robot including sensors to extract key information from the world, controllers to make decisions about how to act given that information and effectors for acting on the world;this chapter will examine strategies for controlling the robot's actions. As previously discussed, the robot's brain makes the decisions regarding how to control the robot to achieve some goal or some task. However, the brain must be wired or programmed to do this.

Typically, low-level control (robot reflex) is hardwired or programmed using feedback control. Hardware control is simple, fast and specialized. Low-level control can also be programmed by using reactive control. Examples of robot tasks or goals that would require low-level control include stable locomotion, monitoring battery level, obstacle avoidance and wall following. For higher-level control (robot reasoning), it is necessary to have some type of control architecture or method for organizing how the robot uses information to act. High-level control is typically implemented in software. Software control is complex, flexible and general. High-level control involves reasoning and decision making. Examples of robot tasks or goals that would require high-level control include navigation, foraging, localization, exploration or mapping. One key difference between writing programs for robots versus for a computer is the need for parallel vs. serial execution [9]. Unlike a computer, a robot seeks more than to arrive at a final answer but rather to achieve some goal or maintain a state while simultaneously avoiding obstacles, navigating, updating its state and maintaining power. Another key difference between a computer program and a robot program are that the robot must take advantage of opportunistic goal realization and graceful degradation in the presence of inaccurate sensor measurements. *Opportunistic goal realization* is when the robot takes advantage of opportunities to achieve a certain state or goal without blindly relying on a predetermined plan. *Graceful degradation* is when the robot program adjusts based upon sensor uncertainty or failure.

Control is also sometimes categorized in terms of the types of tasks: maintenance or achievement. For example, maintenance control would be for an ongoing task such as to remain balanced or avoid collisions. Achievement control would be for a task to accomplish some goal such as navigation to a distinctive place in the environment, localization or to find an object. Control tasks can also be internal such as monitor battery level and external such as move to the recharging station. This chapter will describe some applications of low-level control such as feedback and reactive as well as higher-level control such as deliberative and hybrid. The final architecture described will be behavior-based control which is ideal for a robot program that requires parallel execution, opportunistic goal realization and graceful degradation [9].

3.1 FEEDBACK CONTROL

Feedback or *closed loop control* has been used for many years in industrial systems and is a method for designing a system to achieve, track and/or regulate to a desired state with minimum error. The plant is the system under control, in this case the robot. The error for a feedback control system is based upon finding the difference between the system's desired state and actual state. Steady-state error is the difference between the desired and actual state after the robot has reached the final value and stopped changing. The system's actual state is typically measured with a sensor or transducer and fed back to the input, thus the name feedback. Overshoot is when the system surpasses the setpoint or desired state. Oscillations occur when the robot overshoots or undershoots the desired state with some frequency. The primary purpose of the controller is to minimize the error while reducing oscillations and overshoot. The controller uses the system error to determine an appropriate input to the plant to cause the error to go to zero. Damping is the process of reducing overshoot and/or oscillations. Sometimes it is not possible to select a controller to meet all of the design specifications and there may be some tradeoffs. Figure 3.1 provides the model for a feedback control system. This method of control is typically referred to as the engineering approach to robot control while control architectures use the AI approach [2].

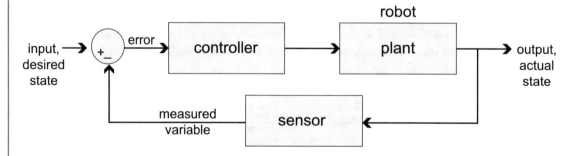

Figure 3.1: Feedback control system model.

The four most common types of controllers are proportional (P), proportional-integral (PI), proportional-derivative(PD) and proportional-integral derivative (PID). Proportional control operates on present values of the system state. Derivative control operates on anticipated future values of the state and adjusts the system momentum or rate. Essentially, derivative control affects the transient response of a system. Integral control operates on the past values of the state and resets that steady-state error. Thus, integral control affects the steady-state response of a system. A controller with a combination of these three components can be used to enable a feedback control system to exhibit desired characteristics and meet design specifications.

Some of these controllers will be examined in the context of creating a wall following robot. The objective will be to use feedback control to design a robot to maintain a distance of 5 in from a wall. If the robot loses the wall, it should randomly wander in the environment until the wall is

detected again. Some ways to create more stringent design characteristics would be to specify the required steady state error, settling time or time to converge on the final value and percent overshoot. Note that open loop control would not be a feasible solution for this type of task because there would be no correction mechanism if the robot drifts too close to or too far away from the wall. Therefore, open loop control is only appropriate for a static environment where the robot and world have a highly accurate, predictable model. The following pseudocode provides a possible closed loop solution.

```
if the sensor detects a wall, measure the error
        if the error is negative,
                turn toward the wall
        if the error is positive,
                turn away from the wall
        move forward
else randomly wander
```

This algorithm is really a state machine where there are four possible states: no wall, too far, too close, just right. This type of system is sometimes referred to as *bang-bang control*. The robot actions include move forward, wander randomly, turn toward the wall, and turn away from the wall. It should be evident, that it will be very difficult for a robot to use this method to follow a wall. Due to sensor error, odometry error including wheel slippage, the robot will not be able to maintain a distance of 5 in. Therefore, the robot will oscillate and overshoot the setpoint and may even eventually become unstable or uncontrollable. However, there are some ways to improve this algorithm including using proportional control.

3.1.1 PROPORTIONAL CONTROL

If proportional control is used to design the wall following robot then the robot will turn by an amount proportional to the magnitude and direction of the error as opposed to just the same angle each time. The proportional gain would be used to scale the wheel speeds or the turn angle. Proportional gain produces a larger response proportional to the distance the system is from the desired state. However, if the gain is too high, the robot may still have overshoot and oscillations. If the gain is too low the robot may never reach the desired state. The form of a proportional controller is

$$u(t) = k_p e(t) , \tag{3.1}$$

where $e(t)$ is the error, k_p is the proportional gain and $u(t)$ is the controller output and plant input. The proportional gain could be tuned to produce acceptable performance for the wall follower. The tuning can be done automatically with a good model or by trial and error. Some ways to improve the proportional control algorithm are to have the robot follow the wall within a dead band region such as 4–6 in. This will reduce the number of times the robot has to turn and thus decrease the oscillations. When the robot is in this window, there will be no adjustments to the turn angle or motor speed. Other ways to improve the performance are to adjust the robot with small angles

proportional to the error, calculate the error and make adjustments more often. The pseudocode for the improved wall follower then becomes

```
if the sensor detects a wall on the left and/or right, measure the error
        if the error is less than 4 in, turn kp*error degrees else
        if the error is greater than 6 in, turn kp*error degrees else
        move forward
else randomly wander
```

In addition to following the wall, the robot may encounter inner and outer corners or doorways. There are a couple of techniques that would help the robot navigate an inner corner. The robot could incrementally make a small turn, back up, drive forward and repeat until it negotiates the corner. Note that this method can be time consuming and not very elegant. Alternately, proportional control could be used to turn the robot with sharper angles proportional to how close it is to the corner. For outer corners, the robot could keep track of its current state and if it abruptly loses the wall, it could make a sharp turn back toward the last known location for the wall to attempt to find it before reverting to random wander. Figure 3.2 provides an example of a robot following a wall using proportional control. It may appear that the robot is not really following the wall at all based upon human standards but it is actually crudely tracking the wall. However, take heart, there are techniques for improving the robot's performance.

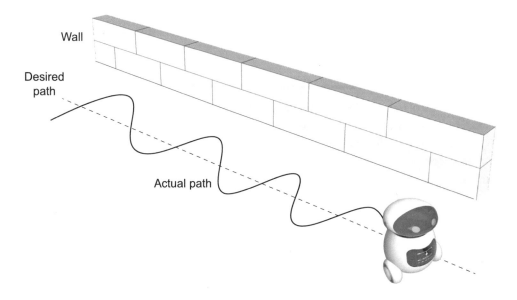

Figure 3.2: Wall following using proportional control.

3.1.2 PROPORTIONAL-DERIVATIVE CONTROL

Proportional control provides a great start for designing a wall following robot, but it is evident that the oscillations are not ideal. The addition of derivative control would be helpful for decreasing the oscillations. Another way of thinking about it is that the derivative control slows down the momentum or rate of change in robot position as it gets close to the desired state. The form of a PD controller is given by

$$u(t) = k_p e(t) + k_d \frac{de(t)}{dt} \, ,$$

(3.2)

where k_d is the derivative controller gain. Since the derivative control is based upon the rate of change, it is necessary to write a program that keeps track of the prior error states and then makes an adjustment proportional to the magnitude and direction of the change in state. Figure 3.3 presents a model of the feedback control system that includes the proportional and derivative controller.

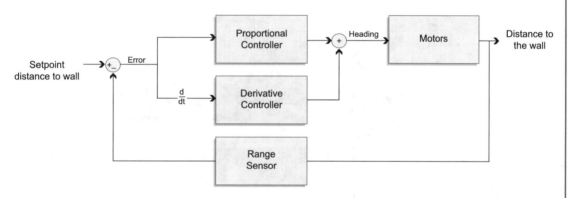

Figure 3.3: PD feedback control.

The key way to reduce the oscillations is to reduce the robot's momentum as it nears the desired state. The application of PD control will adjust the motor power and turn angle proportional to the rate of change of the error. This addition would decrease the sharpness of the turns and reduce the overshoot. PD Control improves the transient or changing response of the system typically with negative feedback. The flow chart in Figure 3.4 describes the improved wall following with PD control.

One disadvantage of this method is that if the brakes are applied too hard or too quickly, the robot may never reach the final state. In addition, although the derivative controller reduces the oscillations, there is a steady-state error. The rate of change of the error is so small that the derivative control cannot correct for it. Also, the magnitude of the error is so small that the proportional control cannot reduce it. Therefore, it is necessary to do something else to make the error zero. Figure 3.5 provides a graphic of a wall following robot using PD control.

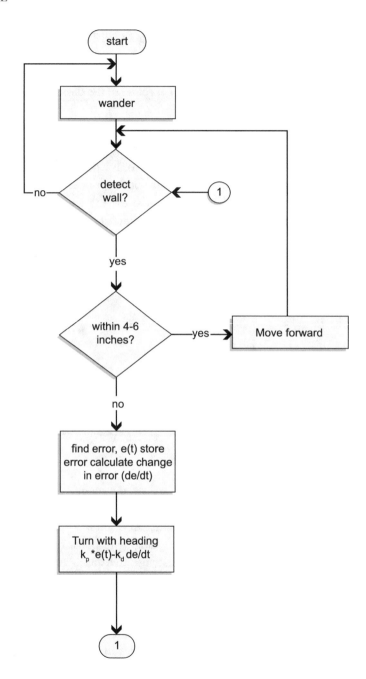

Figure 3.4: PD control flow chart.

Figure 3.5: Wall following using PD control.

3.1.3 PID CONTROL

In order to reset or remove the steady-state error it is necessary to have integral control. An integral controller adjusts the heading of the robot based upon the accumulated error. Typically, the error will have to exceed some threshold before it has a significant effect on the heading. The PID controller has the form

$$u(t) = k_p e(t) + k_i \int e(\lambda)d\lambda + k_d \frac{de(t)}{dt} \, , \tag{3.3}$$

where k_i is the integral controller gain. Figure 3.6 provides the feedback control diagram for the wall following robot.

Typically, the robot's position would be polled at regular intervals based upon the sensor rate and used to find the integral. The integral is the accumulated sum of the error measured at each of these intervals. Similarly, the derivative is the difference between the errors calculated at each of these intervals. Figure 3.7 is an example of a flow chart for the improved wall following algorithm with PID control.

Figure 3.8 presents the improved wall following algorithm with a PID controller. Note that this is only an example and the actual results may be different based upon several variables including the environment, sensor error, robot model, etc. Although this discussion focused on using software to implement PID controllers, it is also possible to build them electronically using operational amplifiers.

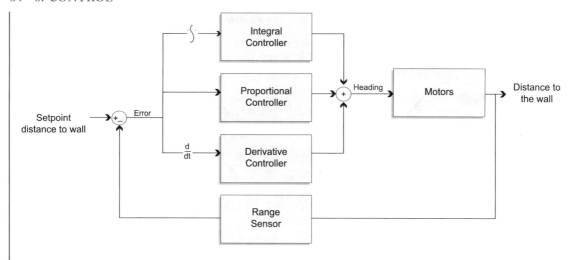

Figure 3.6: PID feedback control diagram.

3.1.4 ZIEGLER-NICHOLS METHOD

As previously stated, it is possible to use trial and error or automatic tuning to set the PID controller gains. However, there is also a method created by Ziegler and Nichols to tune PID controllers that will be discussed here. This method requires adjusting the closed loop gain until the system is marginally stable or exhibits consistent oscillations around the set point. When a system is marginally stable, the oscillations are not damped and ideally could go on forever. In this first step, only the proportional control gain is adjusted, the derivative and integral control gains should be zero. The gain of the proportional controller when the system becomes marginally stable is K_c. The next step is to estimate the period, T_c, of the oscillations in minutes for the marginally stable system. Lastly, use these values and Table 3.1 to solve for the proportional, integral and derivative gains. Note that these values provide a starting part and will most likely need to be tweaked in real time to yield the best performance.

Table 3.1: Controller gains from the Ziegler-Nichols method			
Controller	K_p	K_i	K_d
P	$0.5\ K_c$		
PI	$0.45\ K_c$	$\frac{1.2}{T_c}$	
PID	$0.6\ K_c$	$\frac{2}{T_c}$	$\frac{8}{T_c}$

In conclusion, this has been a very brief summary of how a feedback control system could be used to design a simple wall follower. As the tasks become more complex it will be necessary to

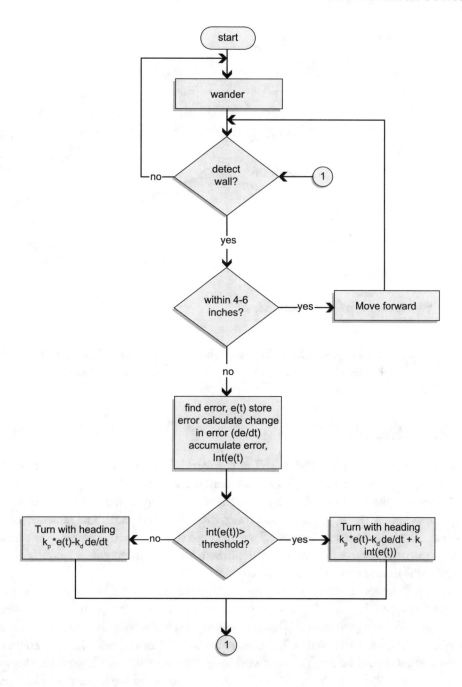

Figure 3.7: PID control flow chart.

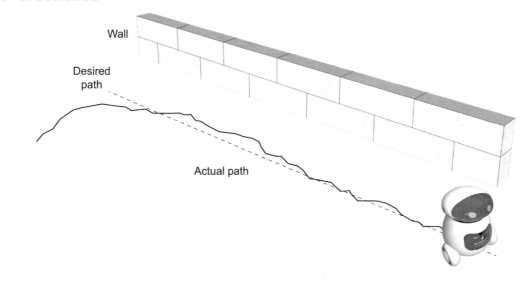

Figure 3.8: Wall following using PID control.

transition from traditional engineering to AI approaches to robot control. Therefore, the next section will address how a robot represents knowledge and then how it is used in control architectures for high-level control.

3.2 REPRESENTATION

Representation is the robot's internal model of the world. This is the information that the robot uses to accomplish goals and tasks. The robot may need to remember past events, use current precepts or predict the future results from its actions. The representation of the robot's self may include proprioception, paths, goals, tasks, sensors, intentions and plans. The representation of the environment may include maps, images of people, objects or distinctive places. For example, the robot can use an odometric path to recall a route traveled. The robot may use a landmark based path to recall salient features and locations in the world.

The robot's representational knowledge is closely related to the ability to sense and extract information from data. Figure 3.9 provides a relationship between the organization of knowledge and the volume or storage memory. This relationship indicates that data produces information and information produces knowledge. Knowledge for robotic control must have an environmental correlation and predictive power. The relationship to the external world must have short term or long term memory, metric or relational mapping as appropriate for the robot's task. The predictive ability of knowledge may be inaccurate or untimely but it is necessary if the robot must do more than purely react to what is sensed.

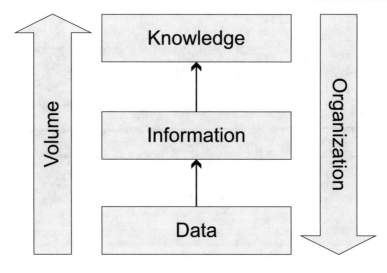

Figure 3.9: Basis of knowledge [10, 11].

Knowledge representation can be explicit, implicit or tacit. ***Explicit knowledge*** is symbolic, discrete and manipulable. An example of explicit knowledge may be the robot's internal model of the world. ***Implicit knowledge*** is not reconstructable but can be made explicit through process. An example of implicit knowledge may be data from a color sensor and data from a laser scanner indicate that there is a red ball in the world. ***Tacit knowledge*** is embedded within the system and cannot be reconstructed from existing processes. An example of tacit knowledge may be a robot's emergent behavior which will be discussed in more detail later in this chapter. Traditional AI robot control systems use explicit knowledge. Modern AI robot control systems use implicit and tacit knowledge.

Knowledge can also be characterized by the time duration. ***Persistent knowledge*** is a priori information that remains static through the robot's execution of a task. Persistent knowledge is the robot's long-term memory. Objects in the world as well as a floor plan may be an example of persistent knowledge. ***Transitory knowledge*** is knowledge acquired during task execution and it is dynamic. An example of transitory knowledge would be other robots in the world or sensor data. Transitory knowledge is like the robot's short-term memory. Figure 3.10 presents a summary of the knowledge time horizon.

The map is the most commonly used world model. An odometric map is based upon the robot's odometry as it travels through the world. This type of map is only useful if the world does not change and the robot is able to accurately track its odometry movements. A landmark based map has distinctive locations such as a light house, golden arches, traffic light and the robot can localize based upon the landmarks. For this map, the odometry does not have to be very accurate but the landmarks must be easily recognizable and static. Landmarks that are linked in a certain order create

Figure 3.10: Knowledge time horizon [10, 11].

a topological map. The robot can use this type of map to navigate and localize. Once again the robot odometry does not have to be very accurate but the landmarks have to be recognizable and static. A map with exact distances and lengths is a metric map and the most likely the most familiar to the reader. This map is the most useful to the robot because of the large amount of information but it requires more storage and the robot must have highly accurate odometry. Long term memory may be represented by a metric or qualitative map while short term memory is represented by a grid-based map. Based upon cognitive science, the increasing levels of intelligence for representing space increase from landmark to topological (or relational) to metric. Figure 3.11 provides a summary of a few of these map categories.

Since the creation and maintenance of a robot's world model requires memory, sensing and computation; different information may have varying timelines. For example, a fully metric map of the environment may last throughout the robot's execution of a task. However, the odometric path may last only until the robot reaches the next via point or the final goal.

3.3 CONTROL ARCHITECTURES

It is evident that feedback control is only applicable to low level control where there is no cognition or reasoning so this section will explore robot control architectures which are necessary for higher level control. A *control architecture* is a set of principles and constraints for organizing a robot's control system [3]. There are three robot primitives used to describe the organization of the robot's control architecture: sense, plan, act. The most prevalent control architectures are deliberative (hierarchical), reactive, hybrid and behavior-based. They differ by the way they handle time, modularity and representation. *Time-scale* refers to how fast the sensing, planning and acting happen. *Modularity*

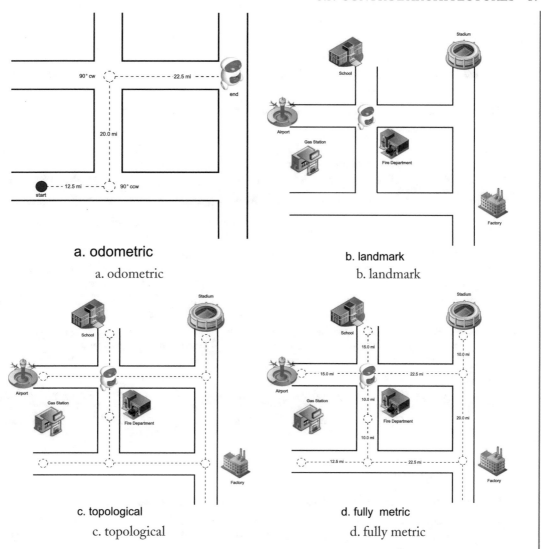

a. odometric

b. landmark

c. topological

d. fully metric

Figure 3.11: Map categories.

describes how the robot primitives interact. **Representation** describes how the robot's knowledge is used.

The deliberative architecture is the oldest control architecture and it is based upon classical AI. It is an expert system which has a top down design and has all of the answers before the task begins. These systems have the longest time scale because they look into the future. Deliberative control has a highly accurate centralized world model. Reactive systems are bottom up and more

adaptive because they achieve tasks by using reflexes. Reactive systems have a shorter time scale because they operate in the present with an immediate response. Reactive control does not facilitate world models. Hybrid control uses a long and short time scale because it includes components of the deliberative and reactive control systems. Behavior-based control brings the time scales together and imposes constraints on how the world models are used. Similar to feedback control, different control architectures are appropriate for different tasks. Table 3.2 provides a summary of the comparison of these four architectures.

Table 3.2: Robot control architecture comparisons [10, 11]				
feature	deliberative	hybrid	behavior-based	reactive
time-scale	long			short
representation	detailed			none
modularity	horizontal	combination	combination	vertical
intelligence	cognitive			low level
action	predictive	combination	combination	reflexive
memory	long term	combination	combination	short term
organization	top down	combination	combination	bottom up

3.3.1 DELIBERATIVE

Deliberative or **hierarchical control** is based upon classical AI. Shakey was the first AI-inspired robot and it used the STRIPS problem solver. The two most famous deliberative control architectures were the Nested Hierarchical Controller by Meystel and the NIST Realtime Control System by Albus. The deliberative paradigm uses a very detailed, accurate world model and considers all of the possible outcomes before taking action. It is a function-based expert system with a horizontal decomposition (top down). It is sequential and orderly; the robot senses the world, constructs a global map, plans all of the actions to achieve a goal based upon the map and then acts to carry out the plan. Note that the robot carries out the plan with no sensing, then senses the consequences of the action and then the cycle starts over. The organization of the deliberative control system is shown Figure 3.12.

Note that the sense may include perception and modeling, the plan may include cognition and localization, and act may include motor commands and motion control.

The advantage of deliberative control is that it provides an ordered relationship between sensing, planning and action. The disadvantage of deliberative control is that it takes a long time to

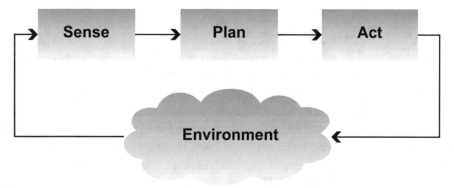

Figure 3.12: Deliberative control organization [8].

form a detailed, accurate world model, search it for the optimal solution and then create the plan as well as update the plan after the robot acts. The problem is that a robot needs a rapid response time to an open dynamic world. The orderly sequential relationship between the robot primitives also means that the sensing and acting are disconnected. To compensate for some of these deficiencies, the robot has to plan rarely and move as much as possible which leads to open loop control. Since there is no verification about how the robot's actions have affected the world, open loop control can be dangerous for dynamic environments. This type of control encourages monolithic programming with niche targetability at the expense of modularity and portability [2]. In addition, the uncertainty in sensor noise and actuator error is not addressed in this model. This means there will be an uncertainty as to whether the action was actually completed. One solution would be fast planning which can be closed loop with more frequent re-planning.

Furthermore, there must be a *closed world assumption* that everything the robot needs to plan to act is available in the representation. However, retaining all of this information can cause a memory shortage or it may not be computationally tractable, i.e., the *frame problem*. Therefore, the environment cannot change during execution, the robot must know the state of the world and the plan at all times. The robot's effectors must also be accurate enough to execute the plan. This horizontal decomposition also makes it difficult to handle multiple goals, multiple sensors, or to have the extensibility to deal with changes in tasks, sensors or locomotion.

Due to these drawbacks, hierarchical control is only appropriate for tasks with no time pressure, detailed advanced planning and a static well-modeled environment. There are not many of these types of applications but some industrial, chess-playing and surgical robots may meet the criteria. Brooks stated that the world was its own best model and this type of planning was a way of avoiding figuring out what to do next, thus reactive control was created [10, 11].

3.3.2 REACTIVE

Reactive control was created in the late 1980s by Rodney Brooks [11] and was modeled after animal behavior. Reactive control is based upon the stimulus response reflex in the context of motor control. It operates very quickly and is effective in a dynamic unstructured environment. In this type of system, there is no complex computation; only short fast pre-computed responses.

Brooks [11] defined a reactive control system as a tight coupling of perception and action to produce a quick response in a dynamic and unstructured real world. This system can be designed by using layers to represent increasing levels of competency. These layers were task-achieving behaviors and used to represent the control system in a vertical decomposition versus the traditional horizontal decomposition of traditional AI control. Reactive control systems are considered to be modern AI control. The motivation for this work was to develop intelligent autonomous mobile robots with multiple sensors, multiple goals, robustness and extensibility. This control system did not afford a world model based upon the belief that the world was its own best model. The lower levels of competence were the class of reactive rules that would be operational in all possible environments (i.e., obstacle avoidance). The higher levels of competence would be specific to a desired class of rules such as reasoning. Arkin [10] stated that the stimulus-response pair was typically modulated by attention and determined by intention. ***Attention*** is used to prioritize tasks and to focus the sensor resources. ***Intention*** is used to determine which rules should be active based upon the robot's goals and objectives.

It is possible to use Braitenberg [13] Vehicles with various sensorimotor connections to demonstrate reactive control. By creating inhibitory or excitatory connections between a robot's motor and a light sensor, the robot can exhibit photophilic or photophobic behaviors. Figure 3.13 shows four possible connections between the robot motors and light sensors. In Figure 3.13 (a), if the right photocell receives more light than the left one, the robot will actually veer to the left away from the source. In Figure 3.13 (b), if the right photocell receives more light than the left one, the robot will actually slow down and move toward the light. In Figure 3.13 (c), if the right photocell receives more light than the left one, the robot will slow down and spend more time in the presence of the light source. In Figure 3.13 (d), if the right photocell receives more light than the left one, the robot will always move in search of more light. Braitenberg referred to each of these behaviors as fear, aggression, love and explorer, respectively.

In a reactive system, each rule typically has a unique stimulus in order to keep it simple and the conditions are mutually exclusive. The complete set of rules for the system is defined at design time. Also to keep the state space tractable, there is a default response for all but the most important events. There are five characteristics of an architecture based upon the reactive paradigm [2]:

- animals are a motivation for the collection of rules

- robots are situated agents

- rules are the building blocks for robot actions and the overall behavior is emergent

- all rules are based upon local ego centric sensing

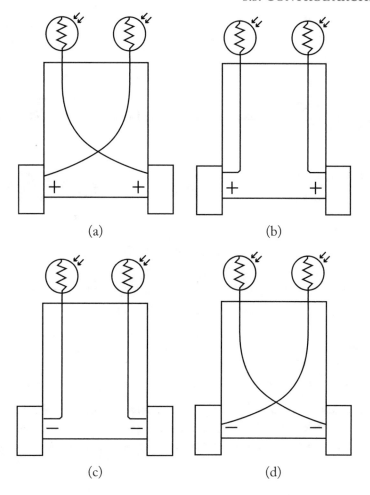

Figure 3.13: Valentino Braitenberg Vehicles [13].

- all tasks can be decomposed into component or primitive rules.

Figure 3.14 is the structural organization of a reactive system in terms of the two robot primitives: sense, and act. Recall that there is no planning in reactive control.

The advantage of programming with reactive rules is that since it is a modular system, each one can be tested independently. In addition, a robot can become more intelligent as more rules are added. The rules can function in parallel with low coupling which affords reuse. The emergent behavior of a robot results from the concurrent execution of multiple rules. In ***emergent behavior*** the robot appears to do more than what it was programmed to do. Some of the limitations of reactive control include minimal state, no memory, no learning and no internal models.

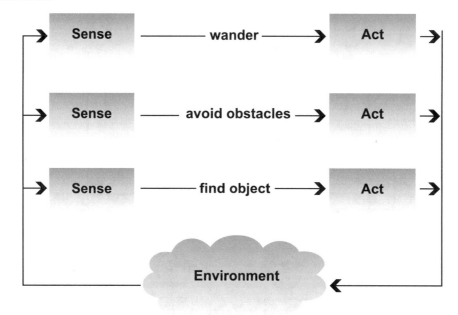

Figure 3.14: Reactive control organization [2, 4].

The two best-known reactive architectures are subsumption and motor schema (potential fields). Subsumption architecture is based upon combining reactive rules in a fixed priority hierarchy. Motor schema is based upon combining the reactive rules as a vector sum for navigation.

Subsumption Architecture

Subsumption architecture was the first form of reactive control and it built incrementally from simple parts to more complex parts. Layers are added from the bottom up with layer 0 being the most simple. Higher layers can subsume lower ones by using them while they are running or by suppressing them selectively. The architecture can be designed to achieve a task by turning the levels on and off. In subsumption architecture, the world is its own best model. In other words, the world can provide all information directly which is better than storing it internally. The components are task-achieving actions or behaviors and the lower layers handle the most basic tasks. The higher layers exploit the existing behaviors. The goal is to have very few connections between different layers. The only connections are inhibition and suppression. There are strongly coupled connections within layers and loosely coupled connections between layers. The input or output of a layer may be inhibited so that it does not receive sensory inputs or send outputs to the motors. Therefore, it will not compute a reaction and send output to effectors or other modules or behaviors. The input or output of a layer may also be suppressed and have the signal replaced by one from a higher level. Note that the

inhibition or suppression may be at either the input or output. The key difference between inhibition and suppression is whether a signal is turned off or replaced by another signal. Figure 3.15 presents one layer of a subsumption architecture.

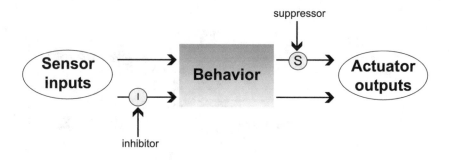

Figure 3.15: Subsumption architecture layer.

The individual layers work on individual goals asynchronously and concurrently. Although not shown in Figure 3.15, this layer is an augmented finite state machine (AFSM) that could also have a reset input to return the behavior to its initial state. There is no global model of the world and each layer has its own perception. The lower levels have no awareness of the higher levels and if the higher levels fail, the lower ones are always functional. Since the complex actions subsume the simpler actions, there is a fixed priority based upon the hierarchy. Thus, even the lowest level of the architecture is a fully functional control system.

Figure 3.16 provides an example of a subsumption architecture with three layers. Layer 0 is obstacle avoidance and there are two reactive rules based upon the sensor input. If the obstacle is directly in front of the robot then the collide behavior will halt or stop the motors. Otherwise, based upon the polar plot generated by the sensors, the robot can feel the force and move away from the obstacle. Layer 1 is the wander behavior which generates a heading for the robot to move. Note that the input to avoid is a combination of the wander and feel force behavior to insure that the robot does not wander into an obstacle. Also note that the avoid behavior will subsume the run away heading in order to insure that the robot uses the modified heading. The highest level of competency is layer 2, homing. The robot will look and if the base station is located, plan a path to it. Note that the heading from the path plan will subsume the heading from the wander. Also, note that the distance and direction that the robot has previously traveled is integrated with the planned path to the base station.

Motor Schema

In 1987, Ron Arkin [12] proposed motor schema based navigation for mobile robots. This method was chosen because path planning and navigation are essentially a collection of behaviors. Recall that a ***schema*** is a basic unit used to describe the interaction between perception and action or for a

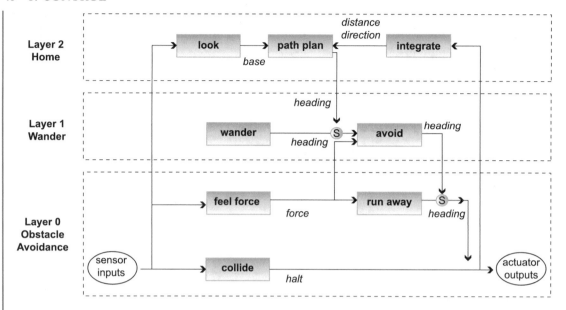

Figure 3.16: Subsumption based robot control system [2, 11].

behavior specification (see Figure 3.13). It is sometimes referred to as a pattern of action as well as a pattern for action. It has knowledge of how to act or perceive as well as the computational process. Schemas provide the behavioral primitives that can be distributed and act in parallel concurrently and independently to create more complex behaviors and allow the robot to move intelligently based upon environmental stimuli.

Schema theory is a method for encoding robot behavior that uses a cooperative method of control. This is in direct contrast to subsumption architecture which uses a fixed hierarchy. Schema theory is typically used for navigation tasks where the sensorimotor mappings are represented by vector-based potential fields to create goal-oriented movements. The coordination of the behaviors is through a cooperative method by using vector addition. Therefore, this type of control is sometimes referred to as the *potential fields method*.

In the potential fields method, the goals are attractors and the obstacles are repulsors. The potential fields are combined through superposition to create a global field that the robot uses to navigate. The vectors represent the behaviors and the vector summation produces an emergent behavior. Each potential is a vector with a magnitude and direction so that the force imparted on the robot is inversely proportional to the distance squared

$$Force \propto \frac{1}{Distance^2}. \tag{3.4}$$

The way to keep the potential field method from becoming an intractable problem is to only compute the instantaneous force where the robot is currently located. For navigation purposes, it is not necessary to create the entire field at once. Table 3.3 provides a summary of some common primitive potential fields.

The potential field represents the motor schema or what the robot should do. This movement is based upon the perceptual schema or what the robot perceives such as an obstacle or goal. Examples of perceptual schema would be find-intersection, find-goal and find-obstacle. Table 3.4 provides a summary of some sample motor schema.

Figure 3.17 provides of an example of a robot using the potential fields method to navigate from a start point denoted by the robot to a goal. Note that this figure is a simplified version of the potential field in order to add clarity. If the robot does not feel the force from any of the vectors then it would not move.

Figure 3.17: Potential fields navigation.

One disadvantage is that this type of control is only appropriate for navigation tasks. In addition, it is vulnerable to local minima such as a cyclic-oscillatory behavior when the vectors sum to zero. There are several ways to handle this phenomenon such as introducing random noise, adaptive methods, time-varying potential fields, harmonic potential fields or timing out to a new behavior such as avoid-past or keep a history of the robot's actions and use that information to make a decision about the next move. Figure 3.18 provides an example of robot navigation where the robot encounters a local minima.

Table 3.3: Primitive potential fields [2, 10]

Type	Graphic
Uniform	
Perpendicular	
Attractor	
Repulsor	
Tangential	

Table 3.4: Motor schema [2, 10]

Motor Schema	Description
Stay-on-path	Robot moves toward the center of a path and moves forward
Avoid-past	Move away from areas the robot has recently visited
Dock	Move toward an object such as a home base in a particular direction
Noise	Moves the robot in a random direction for some period of time
Avoid-obstacle	Move away from an object such as a heat source
Move-to-goal	Move toward a detected goal object such as a charging station

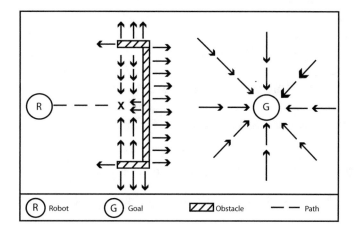

Figure 3.18: Local minima state.

3.3.3 HYBRID

Hybrid control is a combination of reactive and deliberative control architectures. Reactive control is flexible, fast and robust in complex and dynamic unstructured environments and not appropriate for high level control. Reactive control has no planning, memory or reasoning. Deliberative control is appropriate for tasks with a broader perspective and scope and can integrate knowledge. Deliberative control is appropriate for static environments that are easy to model. Therefore, hybrid control was created to deal with limitations in reactive and deliberative control by integrating some planning and knowledge with a purely reactive system. There are three modules or layers in hybrid control including the reactive, planner and middle layer. The organization of the three robot primitives is shown in Figure 3.19.

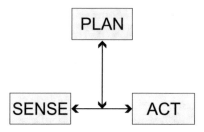

Figure 3.19: Hybrid control robot primitives.

This type of control has the capability of handling different representations, different time scales, and different commands using the middle layer. The middle layer handles behavior management or performance monitoring. It may direct the deliberative layer to perform dynamic replanning based upon the task execution. The reactive layer operates in the present or real time based upon the sensory input. The deliberative layer operates in the past to create plans or a priori maps. The deliberative layer also operates in the future to make projections based upon the results of plan execution. The reactive layer handles local control tasks while the deliberative layer handles global control tasks (i.e., obstacle avoidance versus path planning).

The middle layer has four different techniques for linking real-time reactive rules with long-range planning:

- Selection - the planner can select the behavior to achieve the task

- Advisor - the planner can advise the reactive layer but it is not required for the reactive layer to use it

- Adaptive - the planner can adapt the reactive components based upon the task execution

- Deference - the planner can defer decisions until later in the task execution

Note that the deliberative layer may have real-time or a priori maps and plans for decision making. Figure 3.20 shows the three-layer hybrid control architecture.

Typically, all hybrid control architectures have a sequencer, resource manager, cartographer, mission planner and performance monitor/problem solver. The **sequencer** determines the order in which the behaviors execute to achieve a task. The **resource manager** determines which resources are available for a behavior such as the perceptual schema. The **cartographer** stores and maintains the maps. The **mission planner** determines the overall plan to achieve the task. The **performance monitoring** determines if the robot is making progress toward the goal and may inform the deliberative layer to create a new plan if it is not.

Some of the disadvantages of hybrid control are based upon the fact that the middle layer is difficult to design or debug. It is difficult to determine the division of task responsibilities between

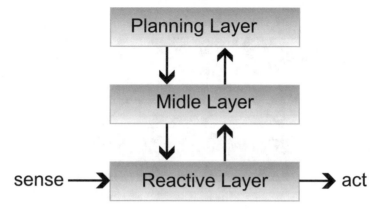

Figure 3.20: Three-layer hybrid control architecture.

the planning and reactive layer. Therefore, the middle layer may have to be custom made for a particular mission or task. In addition, if the middle layer causes the planner to operate too slowly then it may be ignored in task execution.

3.4 BEHAVIOR-BASED CONTROL

Behavior-based control is similar to reactive control because it is fast and reflexive but it also has the capability for learning. The organization of the robot primitives for a BBC looks the same as for reactive control. A ***behavior*** is a close coupling between perception and action. This type of behavior is more expressive than the task-achieving layer used in subsumption architecture. Behaviors achieve and/or maintain particular goals; they are time extended and not instantaneous. A ***ballistic behavior*** achieves a particular goal such as moving to the recharging station or foraging. A ***servo behavior*** maintains a state such as obstacle avoidance or stable locomotion. ***Primitive behaviors*** are composed of only one perceptual schema and one motor schema. The motor schema represents the template for the physical activity and the perceptual schema embodies the sensing. Behaviors assembled from other behaviors are ***abstract behaviors*** and they may have multiple perceptual schema and motor schema. Figure 3.21 shows the behavior and schema relationship.

Behaviors may have different levels of complexity and time but they all can be combined in a behavior-based control. For example, obstacle avoidance would be instantaneous while homing would take a little longer but they can both be operational at the same time. This referred to as bringing the time scales together. Since reactive and behavior-based systems use parallel rules and behaviors, they are both ideal for emergent behavior. Table 3.5 provides a classification of robot behaviors.

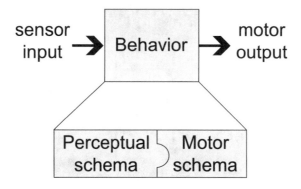

Figure 3.21: Behavior definition [2].

Table 3.5: Robot behavior classifications [10, 11]		
Classification	**Description**	**Behavior (type)**
Exploration	Move in a general direction	Random Wander (servo)
Goal-oriented	Move toward an attractor	Homing (ballistic)
Aversive	Prevent collision	Obstacle avoidance (servo)
Path following	Move on a designated path	Wall following (servo)
Postural	Maintain stability	Balance (servo)
Social/cooperative	Move together	Flocking, swarming (servo)

There are several methods for representing behaviors in a robot control program. A mathematical definition for a behavior or *functional notation* is given by Equation (3.5):

$$b(s) = r \, , \tag{3.5}$$

where b is the behavior, s is the stimulus and r is the response. An example of an obstacle avoidance behavior is given by Equation (3.6):

$$\text{avoid} - \text{obstacle}(\text{detect} - \text{object}) = \text{turn} - \text{motors} \, . \tag{3.6}$$

The benefit of functional notation is that it is straightforward to convert this representation into a computer program. The *finite state acceptor (FSA)* is another useful behavior expression that makes it clear which behaviors are active at any time and the transitions between them. It shows a temporal sequencing of behaviors. Figure 3.22 shows a FSA for an exploration robot with obstacle avoidance. This example has 4 different states: start, wander, avoid, and halt. Halt is the terminal state which means that once the robot enters this state the task is complete. Other means that all other inputs will return the robot to the wander state. There is no terminal state which means the robot will reach a steady-state of going back and forth between avoid and wander. If the robot detects an object it

will avoid and then return to wander and vice versa. The main behavioral state is wander and it may consist of other low level behaviors. This diagram makes it clear the relationships between a set of behaviors.

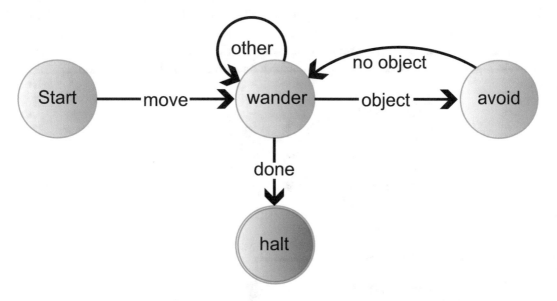

Figure 3.22: Finite state acceptor diagram [12].

A *stimulus-response diagram* can also be used to represent the sequence of behaviors to achieve a particular task. Figure 3.23 provides an example of a stimulus-response diagram. The coordinator is used for behavior arbitration and it will be discussed in more detail later.

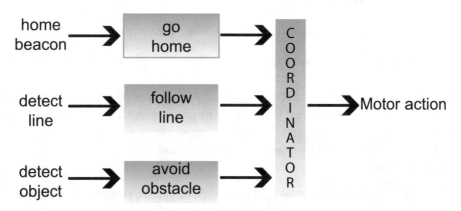

Figure 3.23: Behavior stimulus-response diagram [2].

It is possible to use behavior arbitration or behavior fusion in order to combine behaviors when they are mutually exclusive. **Behavior arbitration** is the method of selecting one behavior to execute. Behavior arbitration is also referred to as **competitive method** where only one behavior can win. The behaviors may have pre-assigned priorities where priorities change at run time. **Behavior fusion** is the method of combining multiple behaviors to execute. Behavior fusion is also referred to as the **cooperative method** because multiple behaviors work together and may create an emergent behavior. Figure 3.24 provides an example of behavior arbitration and behavior fusion.

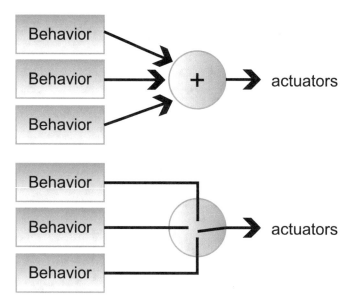

Figure 3.24: Behavior combination [2, 3].

Figure 3.25 provides an example of a behavior-based control architecture.

3.5 REVIEW QUESTIONS

This chapter details several methods to control a mobile robot to achieve a goal or maintain a state. There is a summary of feedback control that is typically used for low-level control as well as control architectures used for high level control. In addition, representation is described as well as how it is used in control architectures. The following questions provide a review of some of the content examined in this chapter.

1. What happens as the proportional gain of a feedback control system is increased?

2. What can you use to reduce the momentum in a feedback control system?

3. Does integral control affect the transient or steady-state characteristics of a system?

Behavior Modules

Figure 3.25: Behavior-based control architecture.

4. Name three techniques that can be used to tune a PID controller.

5. What type of map is appropriate for long-term memory? Short-term memory?

6. What type of map category requires the least amount of memory?

7. What are the three robot primitives? Show the organization of the robot primitives for the four robot control architectures described in this chapter.

8. Control architectures are distinguished by the way they handle what three items?

9. In your own words describe the frame problem and closed world assumption.

10. What are three disadvantages of hierarchical control architecture?

11. Why does deliberative control afford open loop control?

12. What was the motivation for the creation of reactive control?

13. What is the difference between a behavior and a reactive rule?

14. What are the three ways to represent behaviors? Which of them provides a temporal sequence of the behaviors?

15. What type of primitive potential fields could you combine to create a follow-sidewalk behavior?

16. Why would you need to use inhibition and/or suppression in a subsumption architecture?

17. What happens in a subsumption architecture if the higher levels of competency fail?

18. What are three limitations of reactive control? How could you resolve them?

19. Name two ways to extract a robot from a local minima state.

20. Which control architecture is most appropriate for a dynamic, unstructured environment? static environment that is easily modeled?

21. How many layers are typically in a hybrid control system? Which one operates in the present? Which one operates in the past and future?

22. What are the five key components typically present in a hybrid control architecture?

23. In hybrid control, what are the four methods that the planner can use to achieve tasks with the reactive layer?

24. What is the distinct difference between hybrid control and behavior-based control?

25. What is the distinct difference between reactive control and behavior-based control?

26. Would you be more likely to see a system exhibit emergent behavior with behavior arbitration or behavior fusion?

3.6 EXERCISES

1. Use proportional control to design a robot to follow a wall on either side within 4–6 in. The wall following behavior should be rigorous enough to allow the robot to follow inside and outside corners and have a method to extract itself from stuck situations.

2. Improve the wall following algorithm in exercise 1 by adding derivative control.

3. Improve the wall following algorithm in exercise 2 by adding integral control.

4. Improve the wall following algorithm by adding a follow center behavior such that if the robot encounters a wall on both sides it moves to the center of the hallway.

5. Write an algorithm with pseudocode and a flowchart to implement a PI controller to enable a robot to follow a line.

6. Deliberative Exercise. Use a deliberative control to design a robot move to a designated position in the world given the sensor precept. The robot will have a scaled down world model based upon 4 or 5 distinct locations. The robot will use odometry to move to a final location and then repeat the cycle.

7. Reactive Exercise. Use subsumption architecture to create a smart wander routine with the lowest level random wander and the highest level wall following or obstacle avoidance. The robot should randomly wander in the environment until a wall is found and then follow it until it is lost and then revert back to random wander.

8. Hybrid Exercise. Use potential fields to design a robot to home and dock on a heat source or light source. The robot should use a reactive behavior to follow a wall until the source is detected. The robot should then abandon the wall and home on the source, make a 180 degree turn to dock and use the odometric path to return to the wall at the position where it left off and continue following the wall. The robot should create a coarse world model based upon odometry as it traverses the path to the source. The coordinator should arbitrate when the reactive and deliberative control is active.

9. BBC Exercise. Similar to exercise 8, use the robot to home and dock on a heat or light source. The robot should use a wall following behavior until the beacon is detected and then abandon the wall and use a homing behavior to dock on the beacon.

10. Emergent Exercise. Use a temperature or light sensor to implement one of Braitenberg's vehicles on your mobile robot. By wiring each sensor to the motor on the same side or opposite side with a positive correlation the robot will demonstrate fear or aggression, respectively. By wiring each sensor to the motor on the same side or opposite side with a negative correlation the robot will demonstrate love or exploration, respectively.

11. Perform a literature review on the Nested Hierarchical Controller (NHC) and the NIST Realtime Control System (RCS) and compare and contrast their features.

12. Select one of the reactive control research robot platforms from the following list and prepare a research report that describes the control architecture and capabilities. Typically, university websites such as MIT or CMU or Google Scholar provide good references for citations.

Allen

Buzz

Cog

George

Hannibal, Genghis and Attila

HARV

Io, Callisto, Ganymede

Kismet

Polly

Ren and Stimpy

Seymour

Squirt

Tito

Tom and Jerry

Toto

13. Research the following hybrid control architectures and compare and contrast the way they are implemented.

Autonomous Robot Architecture (AuRA)

Atlantis

Dynamic Autonomous Mobile Navigation (DAMN)

Planner-Reactor Adaptation

Procedural Reasoning System (PRS)

Saphira

FX

CHAPTER 4

Software

In the prior chapters, we discussed techniques for controlling mobile robots to complete tasks, achieve goals and maintain certain states. Recall that feedback control was typically implemented in hardware and used for low-level control. However, control architectures were necessary for higher level control and they were typically implemented in software. This section will also discuss software but in the context of navigation competence. From Chapter 3, it was shown that reactive control in particular potential fields were used for low-level navigation. However, reactive control cannot be used for high-level navigation where cognition such as planning, localization and/or map making is required. In other words, global path planning requires localization. This type of navigation requires hybrid control where the deliberative layer takes care of the cognition while the reactive layer performs the path execution. The motion control will include local control strategies such as wall following and obstacle avoidance as well as global control strategies such as finding objects. Figure 4.1 shows a comparison between behavior-based navigation and map-based navigation.

4.1 NAVIGATION

Navigation is the process of purposeful decision making to plan and execute a path from a start position to a goal location. Successful navigation may require localization, searching, path planning, map making, obstacle avoidance, and motion control. *Localization* is the process the robot uses to identify its location on an a priori world map. *Search* is the method the robot uses to efficiently cover the world to locate the goal location. *Path planning* is the identification of a trajectory that will move the robot from a start position to a goal location. *Map making* is the process of using a robot's motion in the environment to create a world map. *Path execution* is the use of reactive behaviors to move the robot from a start position to a goal location. *Obstacle Avoidance* is the modulation of a trajectory in order to avoid collisions as the robot moves to the goal location.

In order to make navigation a computationally tractable problem, the world is typically scaled down from a three-dimensional to two-dimensional representation. In this configuration, the robot is assumed to be a point and the world representation is called the *configuration space (C-space)*. In a C-space, the world is stored in terms of salient features or objects relevant to navigation. Figure 4.2 provides an example of a converting a three-dimensional world to a C-space. There are several types of C-spaces including Voronoi diagrams, visibility graphs, regular grids, and quadtrees.

The robot's world representation is sometimes referred to as the robot's spatial memory. There are four basic functions of spatial memory: attention, reasoning, path planning and updating. *Attention* involves determining what features or landmarks the robot should identify for successful

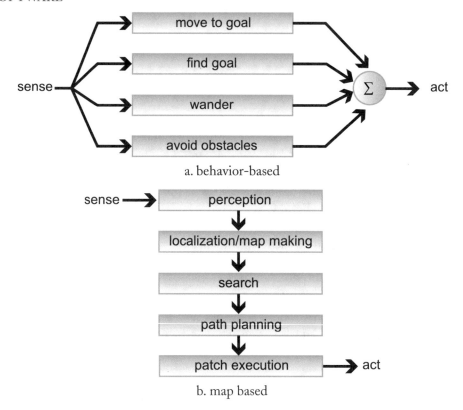

Figure 4.1: Navigation [7].

navigation. ***Reasoning*** addresses the dimensions of the space and whether it is feasible for the robot to navigate through it to the goal. ***Updating*** involves adding new information to the map as well as modifying the plan if necessary as the robot moves about the world. **Path planning** determines the best way for the robot to reach the goal. There are several criteria that a good path planner must satisfy including:

- sufficient complexity to achieve the task

- adequate representation of the terrain

- awareness of the physical limitations of the robot

- compatibility with the reactive layer

- a mechanism to update the map and re-plan [2]

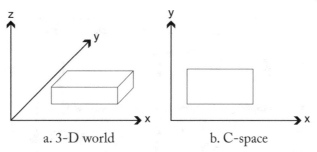

a. 3-D world b. C-space

Figure 4.2: Configuration space.

4.1.1 TOPOLOGICAL NAVIGATION

Topological (or route) navigation is qualitative and involves identifying connections between landmarks in the world. Route navigation is orientation and perspective dependent which is necessary in order for the robot to move from an initial to goal position. A *landmark* represents perceptually distinctive features of interest in the world. A *gateway* is a landmark where the robot has an opportunity to change its overall direction of navigation. A landmark can be either artificial or natural. A *natural landmark* is an existing feature in the world such as a tree or firehouse that can be used for navigation. An *artificial landmark* is an object added to the world to aid navigation such as an arrow or highway sign. There are several important qualities in order for a landmark to be useful including the following:

- easily recognizable

- perceivable from different viewpoints

- supports the navigation task [2]

Route navigation involves turning the world map into a graph with arcs and nodes to make searching for the path from the start point to goal more efficient. Then, the path planner will connect the nodes to create the most optimal path from the beginning to end. Typically, a reactive control strategy will be used to traverse the arcs from node to node.

A *relational graph* has nodes with landmarks, gateways and goal locations with edges that represent the navigable paths. Labeled on the edges are the local control strategies such as follow hallway and move through doorway. Figure 4.3 shows an example of a relational graph.

Table 4.1 presents a transition table for the relational graph in Figure 4.3.

4.1.2 METRIC NAVIGATION

Metric (or layout) navigation is quantitative navigation where the robot's spatial memory is a scaled version of the real world. Layout paths can be used to generate route paths but not vice versa. Layout navigation is a bird's eye view perspective of the world and it is not orientation or position dependent.

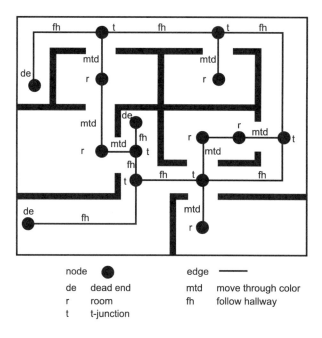

Figure 4.3: Relational graph [2].

	room	t-junction	dead end
	Table 4.1: Relational graph transition table [2]		
room	move through doorway	move through doorway	undefined
t-junction	move through doorway	follow hall	follow hall
dead end	undefined	follow hall	undefined

A search algorithm such as A* or D* is typically used to plan the path for layout navigation but there are also other techniques. Road maps and cell decomposition are typically used for metric navigation.

A *roadmap* is a type of navigation that identifies a set of routes within the free space based upon salient features in the environment. This type of navigation is fast and efficient in spare environments but becomes slower and less efficient as the environment becomes denser. This type of method generates a complete solution which means that there exists a trajectory to the goal for all possible robot poses. There are several types of road maps including visibility graphs and Voronoi diagrams.

Visibility Graphs or Meadow Maps are based upon obstacle geometry where the roads come as close as possible to the obstacles. This method represents the minimum length or optimal path solution but it is at the sacrifice of safety. This technique involves joining all of the edges that can see each other thus the objects in the environment must be polygons. The size of the representation

increases based upon the number of polygons in the space. The free space is then represented as convex polygons and represents the safe region for the robot to traverse. The disadvantages in this type of navigation are that the robot comes as close as possible to obstacles, the trajectories are not smooth and it is also not easy to use sensor data to represent corners and edges for navigation. Some solutions to resolve these problems are to grow the obstacles by the robot's radius, relax paths to create smooth trajectories or move the robot between the midpoints of the segments connecting the corners or in the free space between the line segments. However, all of these solutions will sacrifice the optimality of the solution. Figure 4.4 (a) demonstrates navigation using a visibility graph. Figure 4.4 (b) shows an example of using a visibility graph by connecting the midpoints of the line segments. Note that neither of these examples have path relaxation or the obstacles grown. One disadvantage with growing obstacles is that this method will sometimes close off narrow passageways.

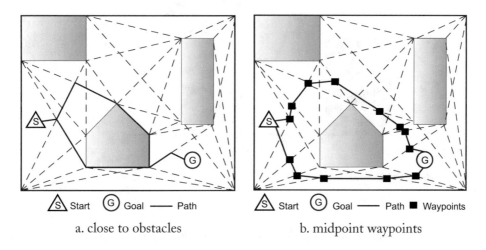

a. close to obstacles b. midpoint waypoints

Figure 4.4: Visibility graphs.

Voronoi Diagrams are also based upon obstacle geometry but the road stays as far away as possible from the obstacles. Voronoi diagrams are also fast and efficient in sparse environment but become less so in denser environments. The Voronoi edge is equidistant from all points on objects in the world. The Voronoi vertex is the point where the Voronoi edges meet. This road map maximizes the distance between the robot and obstacles. This is not the optimal solution for the path length but it is easier to execute based upon local control strategies such as follow center. This method is safer so there is no need to grow obstacles. However, there is a limited range on sensors but if this can be overcome this method is also ideal for automatic mapping. Figure 4.5 shows navigation from a start to goal point using a Voronoi diagram.

Cell Decomposition is a method used to parse a world into free and occupied cells based upon the objects in the world. Then a path plan is generated by creating a connectivity graph through the free cells from the initial to goal state. Lastly, a motion control algorithm such as passing through

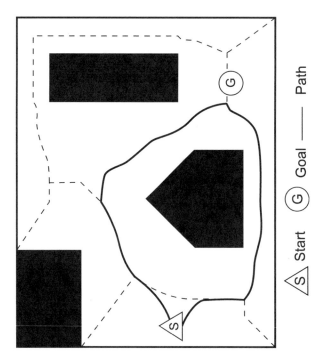

Figure 4.5: Voronoi diagram.

the midpoints of the cell boundaries or the center of the cells is used to traverse to the goal point. There are three types of cell decomposition including exact, approximate and adaptive (quadtree).

Exact cell decomposition is a method for representing free and occupied space where boundaries are placed as a function of the objects in the environment. One technique for doing this is making vertical lines through the corners of the polygons which create convex polygons in the free space. The path planning would involve using the midpoints of the cell boundaries or an arbitrary space in each of the cells. However, this method is rarely used due to the complexities in implementing the path. Figure 4.6 shows an example of navigation using exact cell decomposition.

Approximate cell decomposition is a regular grid representing free and occupied space that approximates the actual world environment. This type of decomposition places a two-dimensional Cartesian grid over the world. It is sometimes also referred to as an *occupancy grid*. This is one of the most popular types of navigation because it is not dependent on objects in the environment and it has low computational complexity. This type of representation will become more complex as the world becomes denser. Another disadvantage of this type of navigation is digitation bias which means that if any part of an object falls in the cell then the entire cell is marked as occupied. Another problem is that this decomposition may sometimes close off narrow passageways.

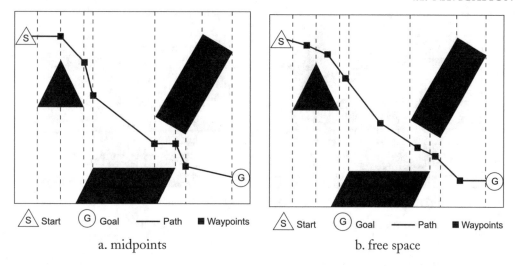

a. midpoints b. free space

Figure 4.6: Exact cell decomposition.

Wavefront or grassfire expansion is typically used to plan a path from the initial to goal position for approximate cell decomposition. However, there is a memory cost from using this type of path planning. Wavefront expansion is the technique of labeling the steps from the goal location back to the start point and then moving the robot incrementally to the goal by counting down at each step. A wavefront can have four-connected or eight-connected neighbors which differ by whether the robot can make diagonal steps. Figure 4.7 (a) presents a navigation example by using a 4-connected wavefront. Note that this method will require 25 steps for the robot to reach the goal. Figure 4.7 (b) presents a navigation example by using an eight-connected wavefront. Note that this method only requires 14 steps.

Adaptive cell decomposition is variable size cell decomposition. The size of the cells will adapt based upon the complexity of the environment. Based upon this modification, the representation will consume dramatically less memory compared to the approximate cell decomposition. Therefore, this type of cell decomposition is one of the most popular. This method is preferable to approximate cell decomposition because narrow passages may not be lost since the cell size is dependent upon objects in the environment and not a regular grid. Figure 4.8 shows an example of navigation using adaptive cell decomposition.

4.1.3 OBSTACLE AVOIDANCE

In the prior chapter, reactive control was used to perform obstacle avoidance in a dynamic, unstructured world. However, with more purposeful navigation that requires cognition it is necessary to have a specific method for avoiding obstacles while continuing to make progress toward the goal. Three techniques to avoid obstacles while continuing to make progress toward the goal are the Bug1,

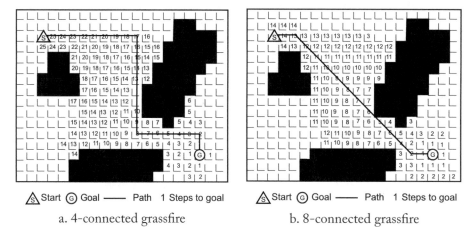

a. 4-connected grassfire b. 8-connected grassfire

Figure 4.7: Approximate cell decomposition.

Bug2 and Tangent Bug. The **Bug1** is the simplest algorithm although not the most efficient because it involves following the contour of an obstacle until it is fully circled and recalling the point when it was the closest to the goal. The robot then returns to that point and leaves the obstacle to continue toward the goal. Figure 4.9 provides an example of the Bug1 algorithm.

The **Bug2** algorithm involves always following the obstacle on the left or right side and leaving the obstacle if a direct connection to the goal location is intersected. This method will have a shorter travel time than Bug1. Bug1 is an exhaustive search algorithm while the Bug2 will move toward the goal at the first opportunity. Figure 4.10 shows an example of obstacle avoidance while moving to a goal by using the Bug2 algorithm.

For the **Tangent Bug** algorithm, the robot will move toward the goal until an obstacle is detected and then follow along the tangential surface of the obstacle until it can once again can move toward the goal. This type of obstacle avoidance approaches the globally optimal path. Figure 4.11 provides an example of obstacle avoidance with the Tangent Bug algorithm.

4.2 LOCALIZATION

The last two tasks in navigation are localization and mapping and they are closely related. These are both open areas of research and difficult problems to solve. They are difficult and closely related because a robot cannot move to a goal point if it does not know its current location in the world. In addition, a robot cannot determine where it is in the world if it does not have a map of the world. Localization is the process a robot uses to determine its location in the world. This is typically done by using odometry and sensor data. Updating the robot's position is separated by the action update form the proprioceptive sensors and the perception update from the exteroceptive sensors. Localization is typically relative to a landmark or reference point in the world and is a **state estimation**

Figure 4.8: Adaptive cell decomposition.

Figure 4.9: Bug1 obstacle avoidance.

problem. Some of the biggest problems with localization are that the estimation process is indirect, sensor measurements may be noisy or incorrect and the appropriate measurements may not always be available. Another problem is that the cognition may require more than just a position; it may need to be with respect to a world model in order to plan a path to a goal.

Belief representation is the robot's belief regarding its current position on the map. This may be a single unique position or a set of possible positions with some type of ranking system. Bayesian is one evidential method that can be used to translate sensor readings into probability functions

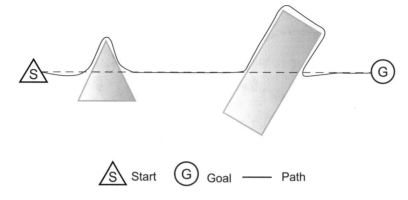

Figure 4.10: Bug2 obstacle avoidance.

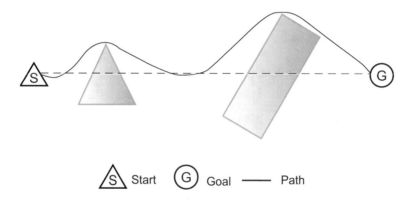

Figure 4.11: Tangent bug obstacle avoidance.

to represent the robot's belief. Markov localization represents the robot's belief by a probability distribution over possible positions by using Bayes' rules. Markov localization allows the process to start from any unknown position in the world. This is possible because the robot can track from multiple, disparate possible positions. This type of localization requires a discrete representation of the space such as geometric or topological. Discrete representation typically has multiple pose estimates and is bound by the resolution of the discretization. In this method, the robot is never lost when the position changes to another cell.

In multiple hypothesis belief, the robot tracks many possible positions and they can be represented as polygons on a continuous or discrete map. Each position can be marked with a confidence or probability parameter. The locations are not ranked unless there is a model for the belief such as a Gaussian probability density function. The advantages of multiple hypothesis belief are that the robot maintains a sense of position while explicitly annotating its uncertainty about the position.

Information from the sensors and effector can be used to update the belief and probability. The disadvantage with this method is that it is difficult for the robot to make a decision without having a single belief. Therefore, this method may prove to be computationally expensive.

Kalman filter localization uses a Gaussian probability density representation to represent a continuous hypothesis belief as a normal distribution. This type of localization uses a continuous map representation which has high accuracy but becomes computationally intractable. Therefore, typically abstraction is used to make the localization achievable. This method does not independently consider each possible robot pose but has a single belief. This method must track the robot from a known position or reference point. It can be used in continuous world representations and is both precise and efficient. The robot can become lost in this type of localization if the uncertainty is too large. Continuous belief representation typically yields a single hypothesis pose estimate and the robot is lost if the position diverges. The advantage of a single hypothesis belief is that it facilitates decision making for path planning. However, this type of position update is challenging and if the robot becomes lost then it cannot perform any cognition.

Kalman filter localization results from the Markov localization if the position uncertainty is assumed to have a Gaussian form. Markov and Kalman filter localization also differ by the perception update. In Markov localization, the entire perception is used to update each possible robot position in the belief state by using Bayes formula. In Kalman filter localization, the perception updates is based upon using the total sensory update to obtain a set of extracted features and then match the features to an object in the map. This method treats the entire unimodal Gaussian belief state at once. The position prediction is based upon the Gaussian error motion model. The observation step is based upon the sensor data that extracts features such as lines or doors. The measurement prediction is the features the robot expects to find at the position. The matching identifies the best pairing between the features extracted during the observation step and the expected features due to the measurement prediction.

There are two types of localization categories: iconic-based, feature-based. ***Iconic-based localization*** uses an occupancy grid and fuses sensor data to correlate with a world model. The fusion is achieved by using by a formal theory of evidence such as Bayesian or Histogrammic in Motion Mapping (HIMM) and is suitable for metric maps. Iconic-based localization uses distances and angles and has a local or relative frame of reference. Relative localization is based upon path integration but this method typically suffers from odometry error. ***Feature-based localization*** uses the Markov decision process and is suitable for topological maps. This type of localization uses the connections among the landmarks and the frame of reference is global or absolute. Absolute localization is based upon proximity to landmarks or beacons.

An example of Markov localization by applying probability theory will be used on the world in Figure 4.12. In the first step, the robot will traverse the world to identify the sensor signature for key features such as distinctive places such as corners (C), hallways (H), dead ends (D) and t-junctions (T). The numbers indicate a gateway where the robot can make a navigation decision. It is possible

to move between the nodes in the world by using local control strategies such as follow center or follow wall to move between nodes.

Figure 4.12: Markov Localization world.

Figure 4.13 provides an example of using the robot sensor model to code the map. Note that this coding could also be used for topological path planning. Once the robot is localized, the topological plan could be something such as left, right, straight, stop. For example, if the robot is placed in the world and moves forward, it will turn left at the first gateway, right at the second gateway, pass the third gateway and stop at the last gateway.

Once the robot is placed in the world, it should use some motion algorithm to move as it keeps track of the gateways passed and the order in which they are passed. At each gateway Bayes rule is used to calculate the probability that the robot is located at a certain position. Bayes rule is a conditional probability given by the following equation to represent the probability of a certain position given a sensor reading:

$$p(A|B) = \frac{p(B|A)p(A)}{p(B)}.$$
(4.1)

Although there will be some odometry error, it would also aid in the localization to keep track of odometry such as distance traveled and turns. Within three to four iterations of this process, the robot should be able to use a Partially Observable Markov Decision Process to localize itself. Figure 4.14 provides a basic example of localization steps for a robot. In step 1, the robot sensor readings indicate that it is in a dead end. Since there are three dead ends in the world, there is a 33% probability that it is in one of 3 places indicated by the robot and the 2 circles. As the robot travels away from the dead end, the robot's probability remains at 33% until it detects the right hallway. At this point, since there are only two places in the world where the robot can move from a dead

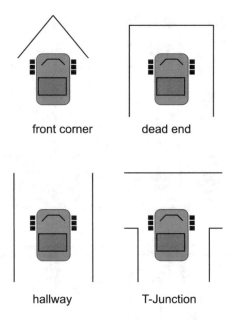

Figure 4.13: Map coding.

end to a right hallway, there is a 50% chance of it being in those two places. However, note that if the robot kept track of odometry then it would be at 100% localized because one hallway is shorter than the other. However, of course this type of localization is at the mercy of the odometry error as the robot moves. This may not be a big concern since this method is also highly dependent on the sensors which also have error. Finally, the robot moves from the hallway to another dead end and that point there is only one place in the world with that sensory sequence so it is 100% localized.

4.3 MAPPING

Mapping is the process of using motion and sensor data to create a model of the world. Mapping requires the robot to learn the environment as well as incorporate new information into the existing world model. This map may also include information about an estimation of the robot's position (localization) and affords cognition so that the robot can do path planning and navigation tasks. Most environments are a mixture of predictable and unpredictable features and the measure of quality is based upon topological and metric correctness. There are several criteria necessary for a map representation to be sufficient for navigation including:

- the level of detail of the map must be appropriate so that the robot can achieve goals

- the precision of the map must be compatible with the robot sensors

a. step 1 b. step 2

c. step 3 d. step 4

Figure 4.14: Markov localization steps.

- the computation complexity of the map is compatible with the computational complexity for reasoning.

For map construction using continuous representation, it is necessary to use two dimensions and represent the objects as polygons to reduce computational complexity. In addition to reduce the memory costs, it is necessary to use the closed-world assumption. This method represents all of the environmental objects on the map in an exact decomposition of the environment. The advantage in the continuous representation is the high level of detail but at the cost of computational memory. One way to make this problem computationally tractable is to use selectivity and abstraction. **Selectivity** is where the map only captures objects that can be detected by the robot's sensors. **Abstraction** is a subset of the features of the real world objects. Another method for simplification is to approx-

imate the real world as a set of infinite lines such as corners and walls. The disadvantage in this method is the lost of fidelity in the world representation.

However, it is possible to create map constructions that are more computationally tractable by using topology or metric information. For a topology-based map, the nodes are put at distinctive locations. For a metric-based, the nodes are put where the features appear or disappear. Topological maps can be quickly constructed and are transient and typically reside in the robot's short-term memory because they are based on the robot's immediate goals. For the cell-decomposition map, the cell boundaries are put where the features appear or disappear in discrete intervals. To create the occupancy grid map representation a counter is used to determine how many times a cell is hit by a ranging sensor. As the counter is incremented, the cell is deemed an obstacle. The memory size increases with the environment size as well as the cell size. This is not compatible with the closed-world assumption which uses small memory requirements with large, sparse environment. Detailed metric maps take a long time to construct and remains in the robot's long term memory.

In order to use Bayes' rule to create an occupancy grid, the first step is to initialize each element in a grid with a probability of 0.5. This indicates that each cell on the grid has equal probability of being occupied or empty. The probability is determined by using conditional probabilities on the sonar model shown in Figure 4.15. In this model, the field of view is β, the maximum range is R, r is the distance to a grid element and α is the angle to a grid element.

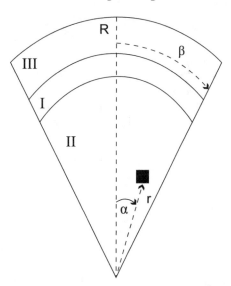

Figure 4.15: Sonar model [2].

Region I represents where objects on the grid may be occupied, Region II represents where objects on the grid may be empty and Region III indicates where the condition of the objects is not known. Obstacles along the acoustic axis of the sonar cone are more likely to be there than

those along either side because of specular reflection. For every element in Region I of the grid, the probability that the cell is occupied is given by the following:

$$P(occupied) = \frac{\left(\frac{R-r}{R}\right) + \left(\frac{\beta - \alpha}{\beta}\right)}{2}. \tag{4.2}$$

Also note that Region I has a finite thickness which represents the tolerance of the sonar. For each element in Region II, the probability that it is empty is given by the following:

$$P(empty) = \frac{\left(\frac{R-r}{R}\right) + \left(\frac{\beta - \alpha}{\beta}\right)}{2}. \tag{4.3}$$

Another method for creating an occupancy grid is by using the Histogrammic in Motion Mapping (HIMM) algorithm. This method was created in order to create an improved and faster obstacle avoidance algorithm. The modified sonar model for the HIMM algorithm is shown in Figure 4.16. Note that this model is only valid along the acoustic axis of the sonar. The uncertainty of the cell being occupied is represented as an integer between 0 and 15.

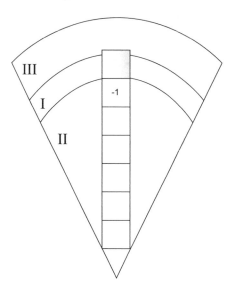

Figure 4.16: HIMM sonar model [2].

The way this method works is that every time a time a sonar reading returns occupied, the cell value for that grid element is incremented by 3. Conversely, every time the sonar reading for a grid element is returned as empty, the value is decremented by 1. These values are capped between 0 and 15. One disadvantage of this method is that if the robot velocity and sensor updates are not well matched then there will be scattered sampling such that a wall may be detected as disparate objects.

The equations used to generate the cell values for the HIMM algorithm are given by the following formulas:

$$grid[i][j] = grid[i][j] + I \tag{4.4}$$

where

$$0 \leq grid[i][j] \leq 15$$

and

$$I = \begin{cases} +3 & \textit{if occupied} \\ -1 & \textit{if empty} \end{cases}.$$

Figure 4.17 shows an example of using the HIMM algorithm to create an occupancy grid. If we assume that a cell value of 15 indicates an occupied cell on the grid, then there are no occupied cells on the map.

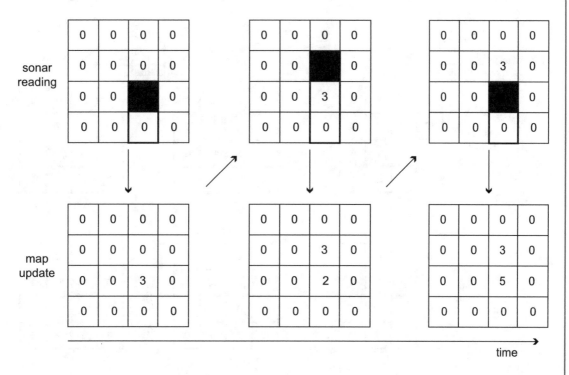

Figure 4.17: HIMM map making.

One way to resolve the problem with the scattered sampling is to use a mask which is referred to as a Growth Rate Operator (GRO). This mask has 3 x 3 dimensions where the multiplier on the occupied cell is a 1.0 and the multiplier on the 8 neighbors is 0.5. One problem is that by resolving the scattered sampling problem, the GRO may produce maps with phantom obstacles. This may

not be a problem as long as the robot is not operating in a highly dense world. The way the GRO works is that if an element has neighbors that are occupied then the score for that element will be increased. The GRO is the 8-connected neighbors around the cell or a 3 x 3 mask and based upon the values on the neighbors it will increase that cell's value. The equations used to generate cell values for the HIMM algorithm with GRO are given by the following two steps:

$$grid[i][j] = grid[i][j] + I \tag{4.5}$$

$$grid[i][j] = \sum_{p,q=-1\ldots1} grid[i][j] \times W[i+p][j+q]. \tag{4.6}$$

Figure 4.18 shows an example of using HIMM with the GRO to create an occupancy grid. If we assume that grid cells with the value of 15 represent occupied cells then there are two occupied cells on the map; one at row 2 column 3 and one at row 3 column 3.

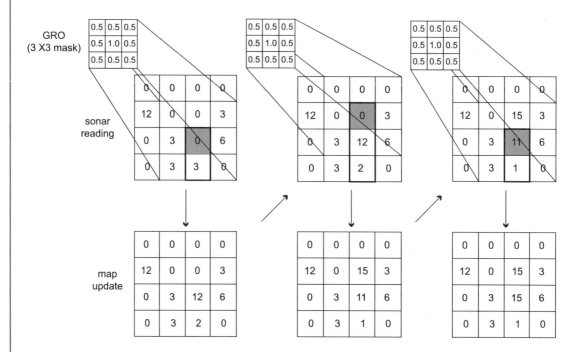

Figure 4.18: HIMM with GRO map making.

In summary, world models require construction, maintenance and computation and they are extremely important to navigation. However, localization is equally important and the combination of the two of them enables a robot to plan a path and move from an initial to a goal state.

4.4 SIMULTANEOUS LOCALIZATION AND MAPPING

Simultaneous localization and mapping (SLAM) is a robot's method of using motion and sensor updates to build a world map and then localize itself relative to the map created. This type of method is also referred to as concurrent mapping and localization (CML). This is one of the most difficult problems in robotics to solve. This is because it is based upon position updates and sensor data which are both error prone and noisy. This is a data association problem because it requires correlation between the sensor and odometry with the world map which is the ground truth although it is also under construction at the same time. This is difficult because the robot has to deal with places that look similar in the world such as landmarks. In addition, the robot has to deal with odometry error and sensor measurements and create a correlation between them. Some of the problems occur if the robot localizes based upon an incorrect feature which then becomes a part of the map. In order to achieve SLAM the robot must cover the world which is assumed to be closed for tractability.

Coverage is the process of searching all of the navigable spaces in the world to create a world map, localize or navigate to a goal point. When the robot does not have a map it has to move in a systematic fashion in the world. Some examples of movement would be wall following, line following, follow hallway, spiral out from the center or random wander. A Generalized Voronoi Graph where the robot attempts to maximize its distance from all objects in the world would be a another systematic and efficient method of covering the world. Figure 4.19 shows an example of creating a world map by using HIMM and a Voronoi diagram or follow center behavior.

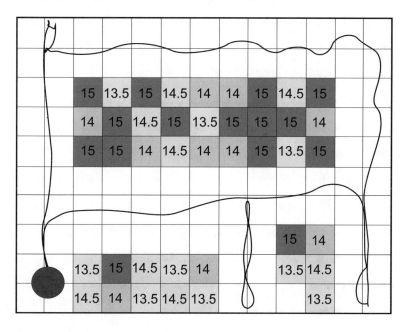

Figure 4.19: HIMM map making.

4.5 REVIEW QUESTIONS

This chapter has summarized several navigation competencies including path planning, mapping and localization.

1. List five competences of navigation.

2. What is the purpose of using a configuration space for world representation?

3. List four types of configuration spaces.

4. What are the four basic functions of spatial memory?

5. Which type of navigation includes finding connections between landmarks?

6. What is the difference between a landmark and a gateway?

7. What type of navigation includes creating a world with road maps?

8. Compare and contrast visibility graphs and Voronoi diagrams.

9. What type of motion control is typically used on the edges on a relational graph?

10. Name three ways to improve some of the problems with visibility graphs.

11. Which type of cell decomposition is the most popular?

12. Which type of cell decomposition is the most computationally intensive?

13. Which obstacle avoidance algorithm is the closest to the optimal solution?

14. What are some of the biggest problems with localization?

15. Would you use Markov or Kalman filter localization on a continuous map?

16. Can a robot recover from a lost position using Markov or Kalman filter localization?

17. What are the two localization categories?

18. What is the difference between selectivity and abstraction in map construction?

19. Which maps typically reside in a robot's short-term memory and are quicker to construct?

20. Which mapping algorithm is only valid along the acoustic axis of the sonar?

21. Is it possible to use any range sensor for the HIMM algorithm? Why or why not?

22. What is one benefit and one disadvantage for adding the GRO to the HIMM algorithm?

23. List at least two reasons why simultaneous localization and mapping is a difficult problem to solve.

24. List at least three different methods for covering and searching world model.

4.6 EXERCISES

1. Use a text file to create an occupancy grid of an environment. Next, use wavefront propagation to write an algorithm to move a robot from a start position to a goal position.

2. Use a text file to create a topological map of an environment where the salient features are wall locations. Next, use reactive control with local control strategies to move a robot from a start position to a goal position.

3. Write an algorithm to create an occupancy grid of the robot's environment using a Bayesian or Histogrammic in Motion Mapping techniques.

4. Write an algorithm to use a robot to create generalized Voronoi graph of an environment.

5. Use a text file to create a qualitative or quantitative map of a robot's environment. Next, use a motion algorithm such as random wander, Voronoi graph or local control strategy to identify gateways until the robot is successfully able to localize in the world.

6. Use topological navigation with local control strategies to navigate a robot through a hallway and into a predetermined classroom by using a state machine.

7. Use MATLAB to create a simulation of a Bug1, Bug2 and Tangent Bug obstacle avoidance algorithms.

8. Program a robot to move to a goal position by using one of the bug obstacle avoidance algorithms.

9. Create a research report on the Rhino, Minerva, Xavier and SAGE tour guide robots and how they performed navigation.

10. Write a research report to describe the localization scheme that the robot, Dervish, used to win the 1994 American Association for Artificial Intelligence (AAAI) .

11. Write a research report on the localization scheme used on the robot, Rhino, by Burgard.

12. Write a research report to compare and contrast the classical potential field and extended potential field methods for navigation.

13. Write a research report on the localization scheme used by the Pygmalion robot at the EPFL.

Bibliography

[1] G. Dudek and M. Jenkin. *Computational Principles of Mobile Robotics*, Cambridge: Cambridge University Press, 2010. Cited on page(s) 1

[2] J. Borenstein. "Measurement and correction of systematics odometry errors in mobile robots", *IEEE Transactions on robotics and automation*, vol. 12, no.6, 1996. DOI: 10.1109/70.544770 Cited on page(s) 3, 4, 12, 28, 41, 42, 44, 46, 48, 49, 52, 53, 54, 60, 61, 62, 73, 74

[3] J. Borenstein, H.R. Everett, L. Feng, and D. Wehe. "Where am I? Sensors and Methods for Mobile Robot Positioning - Sensors and Techniques", *Journal of Robotics Systems, Special Issue on Mobile Robots*, vol. 13, no. 4, pp. 231–249, 1996. Cited on page(s) 4, 38, 54

[4] J.L. Jones, A.M. Flynn, and B.A. Seiger. *Mobile Robots: Inspiration to Implementation*, 2nd edition, Natick: A.K. Peters, 1999. Cited on page(s) 5, 12, 25, 44

[5] J.L. Jones. *Robot Programming: A Practical Guide to Behavior-Based Robotics*, Chicago: McGraw-Hill, 2004. Cited on page(s) 13

[6] M. Boutell, C. Berry, D. Fisher, D., and S. Chenoweth. "A Multidisciplinary Robotics Minor", *Computers in Education Journal*, vol. 1, no. 3, pp. 67 - 80, 2010. Cited on page(s) 13

[7] M.J. Matarić. *The Robotics Primer*, Cambridge: The MIT Press, 2007. Cited on page(s) 17, 60

[8] R. Arkin. "Motor Schema Based Mobile Robot Navigation: An approach to programming by behavior", *Proceedings of IEEE International Conference on Robotics and Automation, pp. 264 - 271*, March 1987. DOI: 10.1109/ROBOT.1987.1088037 Cited on page(s) 16, 41

[9] R. Siegwart and I.R. Nourbakhsh. *Introduction to Autonomous Mobile Robots*, Cambridge: The MIT Press, 2004. Cited on page(s) 27

[10] R.A. Brooks. "A robust layered control system for a mobile robot", *IEEE Journal of Robotics and Automation*, vol. 2, no. 1,pp. 14–23, 1986. DOI: 10.1109/JRA.1986.1087032 Cited on page(s) 37, 38, 40, 41, 42, 48, 49, 52

[11] R.C. Arkin. *Behavior-Based Robotics*, Cambridge: The MIT Press, 1998. Cited on page(s) 37, 38, 40, 41, 42, 46, 52

[12] R.R. Murphy. *Introduction to AI Robotics*, Cambridge: The MIT Press, 2000. Cited on page(s) 45, 53

[13] V. Braitenberg. *Vehicles: Experiments in Synthetic Psychology*, Cambridge: The MIT Press, 1984. Cited on page(s) 42, 43

Author's Biography

CARLOTTA A. BERRY

Dr. Carlotta A. Berry is an associate professor of electrical and computer engineering at Rose-Hulman Institute of Technology. She received her Ph.D. from Vanderbilt University and her thesis was on the development of an enhanced human-robot interface for a mobile robot. She has a Master's degree is from Wayne State University, and holds two bachelor's degrees: one in mathematics from Spelman College and one in electrical engineering from Georgia Institute of Technology. She has been an active member of the American Society of Engineering Education (ASEE) and the Institute of Electrical and Electronics Engineers (IEEE) since 2004. Her research interests include multidisciplinary educational robotics, human-robot interfaces and identifying strategies to increase enrollment and retention of women and minorities in engineering. She is the co-founder of the first multidisciplinary minor in robotics at Rose-Hulman, and she is also co-PI of the Rose-Hulman building undergraduate diversity (ROSE-BUD) program funded by an NSF S-STEM grant to increase the recruitment, retention and development of underrepresented populations in electrical and computer engineering.

Printed in the United States
by Baker & Taylor Publisher Services